COPYRIGHT and PHOTOCOPYING

PAPERS ON PROBLEMS AND SOLUTIONS, DESIGN FOR A CLEARINGHOUSE, AND A BIBLIOGRAPHY

edited by

Laurence B. Heilprin

Student Contribution Series

Number 10

College of Library and Information Services

University of Maryland

1977

Library of Congress Catalog Card No.: 77-620028
ISBN 911808-14-0

TABLE OF CONTENTS

* see Title pages for individual contributors.

PREFACE

The fact that it took almost seventy years to achieve a substantial revision of our copyright statute is sufficient testimony for the assertion that the problems involved are not easy to solve. Even now, with the revision on the statute books, there is dissatisfaction and anxiety among the interested parties.

It may be useful under the circumstances to take a momentary step back from the legal and procedural issues and confront what might be characterized as meta-problems.

The first of the meta-problems has become almost a cliché: to wit; in contemporary human affairs, events tend to outpace plans. Specifically, in the present case, the deliberations of legislatures and their supportive institutions have been outpaced by technological events. Thus, the ultimate solution of the copyright problem may be contingent on the solution of the meta-problem which can be restated as the achievement of some reasonable synchrony between the pace of policy deliberations and the pace of technological development.

A second meta-problem has to do with the variation in the real value systems of the originator-owners of intellectual property.

Put more concretely; widespread, free dissemination can be very good for some originator-owners and can be very bad for others. Thus any rule which either consistently restricts or consistently facilitates photo-copying will not have a consistent value effect on all copyright holders. It would appear, then, that the traditional linkage in legal philosophy between consistency and equity is not entirely appropriate in the copyright matter. Perhaps we need a mechanism whereby an originator-owner can specify how intensively any restrictions should apply to a given work.

Both meta-problems are approachable from a base which draws on new concepts of organizational coupling. Put simply, there are some transactions that seem to require some form of flexible buffer.

The work which follows focuses on the central issue of the form of organization best suited to cope with the varying needs of copyright holders and users; in particular, on the issue of a Copyright Clearinghouse. It is quite conceivable that such an organizational entity could function as the kind of flexible buffer needed to reconcile the needs and values of the originator-owners *vis a vis* the users of intellectual property. Such a proposition deserves rigorous consideration. That consideration has been well begun is demonstrated in the studies which follow.

J. S. Kidd

Acting Dean

INTRODUCTION

CONTRIBUTIONS TO SOLUTION OF THE COPYRIGHT PHOTOCOPYING PROBLEM

This volume presents two reports and a bibliography on the copyright "photocopying problem." The phrase has come to mean a broad set of problems connected with use of the technology of copying a copyrighted work (by photography, xerography, etc.), without specific permission from the copyright holder. In many cases it is in the public interest that such use be made of the work; in others, it is in the public interest to protect the copyright holder against infringement. The two reports represent the combined teamwork of two successive graduate classes in library operations research I conducted at the University of Maryland, College of Library and Information Services, in 1975 and 1976. The annotated bibliography was carried out as an independent study by a student in the 1976 class.

The copyright photocopying problem is one of the most intractible encountered by the post-industrial society--the problem of rights to, and use of, intellectual property in a technological environment which has changed from the simple one-medium of Gutenberg to the multimedia of today. Is it feasible to continue to uphold the Constitutional provision: The Congress shall have the Power to promote the Progress of Science and the useful Arts, by securing for limited Times to Authors and Inventors the exclusive Right to their respective Writings and Discoveries" (Art. 1, Sec. 8(8))? With ever greater ease of recording and copying copyrighted works on the part of a user of these works, it appears less and less effective, or possible, to protect the copyright owner by enforcing his exclusive rights. For the last three decades those concerned with copyright--authors, publishing industries, educational institutions and libraries--also the Copyright Office (charged with administering the Copyright Statute that implements the Constitutional provision)--have been struggling to reach agreement on an equitable means of <u>both</u> protecting ownership of intellectual property <u>and</u> allowing the greatest freedom in technological use of that property by those conducting Science and the useful Arts. Without support for ownership much of the incentive for disclosing discoveries, incorporating knowledge in texts, writing literature, producing works of art and articles of beauty would, under our system of free enterprise, disappear. Somehow, we must maintain a fair return for this effort, the benefits of which many now take freely. Freedom to copy should also be preserved, but access to copyrighted works should be politically free, not economically free, in the sense of costing nothing. Therefore, there is need for such studies as these directed at both adequate return and "free" access to works, using modern competitive technology of copying and transfer. Both principles are valid. The problem is, how to reconcile them? Reconciliation is indispensible to the progress of this information-oriented nation, as well as to that of the global community that looks to this one.

The first study, conducted in 1975, attempted to sift all the main solutions known or proposed, using an independently selected set of criteria for solutions. These were originally chosen by the Committee to Investigate Copyright Problems Affecting Communication in Science and Education (CICP) in their 1959-60 report (published 1962). In 1975 the class reviewed the original criteria, deleted some, added others. Basing the format of their work on that of the CICP report they produced an improved, modern version. This was looked over (as was the original CICP report) by obliging, knowledgeable members of the Copyright Office. Thus, while the actual authors of both reports remained fully responsible, they attempted not to take sides, and to maintain objectivity.

The Copyright Revision Act of 1976 accomplished something for both producers and users. One of the gains for users is that the law now permits freer access for users of copyrighted works, by officially recognizing fair use as a legislative principle. This weakens the exclusive rights of authors and copyright holders. On the other hand much protection of copyright-holders' rights remains, and the law also limits fair use. The limitations, however, require extra time and effort on the part of users and their agents. They impose a need to discriminate the purpose for which copying is done--for profit or not for profit--; to count numbers of copies from a given serial publication made over extended periods; and to avoid a systematic as against a random pattern of use. Another gain for users and loss of exclusive rights is provision for "compulsory licensing" in certain cases. To make compulsory licensing adaptable a whole new system of Federal rate-making is introduced through a Copyright Tribunal. Thus the cost of preserving the copyright principle has been increased by more record-keeping by libraries and other users and by the Copyright Office, by more study by the CONTU, and by study and adjudication by the Tribunal. These "overhead" costs are paid for mainly by the users of copyright and the taxpayers. The copyright holders also pay through diminution of their exclusive rights. It is not clear that such a dynamic dissemination system with its increased "frictional drag" is best for this Nation, or that such a cumbersome system can in the long run survive. There is room and need for a simpler solution.

The analysis contained in the CICP report of 1962 had shown a slight but distinct advantage in favor of a solution called the "copyright clearinghouse." For some years, the CICP Study Group under my chairmanship conducted a number of exploratory investigations, and supplied Congressional hearing testimony, on the nature of such a copyright clearinghouse. Later the CICP, under encouragement from the U.S. Office of Education, in 1967-1968 actually organized a limited copyright clearinghouse experiment. A contract was negotiated between CICP and a large non-profit publisher (American Society for Testing Materials) which in turn contracted with its clients, allowing them to make up to 50 copies free of royalty charge. In return, the clients agreed to supply the CICP with data on their copying experience, for study and planning. Operation of this contract was interrupted by the Williams and Wilkins vs. U.S. case, and termination of the activities of the CICP (Science, Vol. 160, 21 June 1968, 1324-5). Nevertheless the attempt showed some positive results (see Clearinghouse article, ref. 8). In the case of musical works, ASCAP and BMI have successfully demonstrated for-profit clearinghouses under publisher control.

In the revised survey conducted by the 1975 class the clearinghouse was again rated high but there were a number of equally rated alternatives. When the class of 1976 reconsidered the clearinghouse and decided to study it in depth they used my notes on the material of the CICP Study Group, and an illustrative quantitative OR model I began in 1962. The resulting design essentially completes that started in 1962, so that our second report began to take on more of the character of a continuation of the CICP studies than I had intended. My excuse must be that the subject of a copyright clearinghouse, almost dormant for a decade and a half, was now being actively reviewed in many quarters. The Copyright Revision Act of 19 October 1976 has, if anything, increased interest in the feasibility of a copyright clearinghouse. A number of clearinghouse studies or models are now (1977) in process. To my knowlege, however, no economically self-sustaining non-profit clearinghouse for general works, with control as proposed here (sampling rather than detailed accounting; Board of Directors balanced between adversary interests, other public interests represented) has been or is being developed. Our conclusion that a self-supporting copyright clearinghouse depends primarily on volume, and is only now becoming viable, seems to be confirmed by recent events.

The present publication brings the two studies together in such a way that each adds perspective to the other. They are further strengthened by two annotated bibliographies. This first is a selection from large, computer-based national collections of abstracts of articles bearing on the copyright photocopying problem, 1970-1976. The second, a subset of the first, documents the Williams and Wilkins case.

As the reader can imagine, it is scarcely possible for a team of half a dozen graduate students and one professor to write an exhaustive work within the narrow fifteen weeks of one College Term. The main value of each study, then, must be in pointing out a good many considerations until now not understood, or if understood, not publicly stated. In particular, the clearinghouse solution to the photocopying and similar copyright dissemination problems could serve as a preliminary blueprint for a more careful study. Other ideas on D-publishing and D-libraries are adjoined, and the whole area seems a fertile one for investigation of a new, comparatively untouched part of information science.

The coincidence that I as faculty member happened to have taken a part in copyright studies prior to teaching at Maryland accounts for the choice of problem and much of the treatment. This fact should not, however, obscure new insights and other valuable contributions by members of each team. These were substantial. Excellent team work extended over a wide range--from gathering references to study, analysis and formulation of issues; outlining reports; writing sections, critically analyzing the drafts; editing, drawing figures, and making final copy ready for typing. The team members were "turned on"--putting forth more effort and investing more than the nominal three hours per week for course credit.

Although each report separately thanks Barbara Ringer and other members of the Copyright Office, I should like to add my personal thanks and gratitude for their continued intellectual support of the tasks undertaken in these graduate classes.

Before publishing this work the Publications Committee of the College of

Library and Information Services consulted Ben H. Weil of Exxon Research and Engineering. I wish to thank him for some very constructive suggestions used in this introduction.

Last, I should like to acknowledge the help of Esther Herman, Committee editor, in facilitating this publication; Jan Schlesiger and others of the CLIS office staff for its faithful preparation.

Laurence B. Heilprin
Professor Emeritus, Information Science
College of Library and Information Services
University of Maryland

May 10, 1977

PROPOSED SOLUTIONS TO THE PROBLEM OF PHOTOCOPYING COPYRIGHTED MATERIAL: A SURVEY

Report by the Class in Operations Research for Library and Information Science: LBSC 858 H

College of Library and Information Services
University of Maryland
College Park, Maryland

30 April, 1975

GRADUATE STUDENTS

Barbara A. Broady, Robert M. Crowder,
Clodagh M. Lee, Dennis D. McDonald
Betty J. Manchak, Joan P. Mavity,
Kathleen S. Reif, Toba L. Singer,
Michael L. Williams

FACULTY ADVISOR

Laurence B. Heilprin

TABLE OF CONTENTS

ABSTRACT

The College of Library and Information Services, University of Maryland, class in Operations Research for Library and Information Science, with assistance from the United States Copyright Office, Library of Congress, examined the photocopying/copyright problem. Possible solutions were tested against a set of selected criteria. Further evaluation and conclusion are left to the reader.

I. INTRODUCTION and BACKGROUND

This report is the result of a class in Operations Research conducted by Dr. Laurence B. Heilprin at the College of Library and Information Services at the University of Maryland, College Park, Maryland in the Spring of 1975. The class was numbered 858-H and from this the group took its name.

Although the members of the study group were all advanced degree candidates in Library Science, the perspective of the study was not exclusively that of the librarian. It encompassed a range of disciplines including Operations Research, Information Science, and Cybernetics as well as Library Science.

The area of study selected was copyright and photoduplication. Librarians in particular have become increasingly aware of the implications of copyright legislation for information dissemination. However, librarians are far from the only interested parties and it was the hope of the 858-H group that the larger issue of information dissemination through photoduplication could be addressed apart from any specific reference to the implications for libraries.

The May 1960 report of the Committee to Investigate Problems Affecting Communication in Science and Education (CICP) served as both a framework and a "jumping-off" point for the work of the 858-H group. Our intention was to objectively analyze the main aspects of the copyright photocopying problem as we see them and present possible solutions in an unbiased way.

Whereas the CICP report is specific to science and education, the 858-H group has studied the problem with a view to providing solutions which are sufficiently general so as not to be restricted to any particular subset of knowledge.

There has been much activity recently in the field of copyright. The latest version of the Copyright Revision Bill (Senate Bill S. 22 of the 94th Congress, 1st session), the split decision of the Supreme Court in Williams &

Wilkins Co., vs. U.S., and the recently established Commission on New Technological Uses of Copyrighted Works demonstrate the urgency of finding a solution to what many regard as the dilemma of copyright and photoduplication.

It is in this atmosphere of accelerating developments in the field that this study was undertaken and that this report was prepared. Hopefully, the 858-H group has helped to shed some light on what has been a rather dimly lit corner of our field and added impetus to the search for a solution.

The class wishes to express their warm thanks to the following persons for their generous contributions in time, lecture and assistance: the Register of Copyright, Barbara Ringer; the Deputy Register of Copyright, L. Clark Hamilton; the General Counsel, Dorothy M. Schrader; the Chief of Reference Division, Waldo H. Moore; the Librarian Attorney Advisor, Benjamin W. Rudd; and Attorney Advisor, Wilma S. Davis.

At the same time, while acknowledging major help without which the collaborative effort could scarcely have succeeded, the class takes sole responsibility for this report, its results and imperfections.

II. ASSUMPTIONS

The following are the most important assumptions which guided our analysis:

1. The Constitutional copyright principle---"The Congress shall have the Power to promote the Progress of Science and the Useful Arts"---will be retained. The means to accomplish this---"...by securing for limited times to Authors and Inventors the exclusive right to their respective Writings and Discoveries"---will also be retained.

2. A practical result of the assignment and protection of copyright ownership is the economic stimulus given to the creation and dissemination of information.

3. Any proposed solution to copyright revision must recognize both the protection of the copyright owner's rights and economic stimulus for creative works.

4. As a result of advances in technology, it is no longer possible for the copyright owner to prevent duplication and multiplication of copyrighted works. Therefore, the financial return to the copyright owner should include a return for a number of copies made of the work.

5. The capability of a copyright system to promote the generation and dissemination of information depends upon its ability to regulate some elements of the information transfer process.

6. Regulation implies that some mechanism be developed to act as a regulator. Regulation includes both monitoring and controlling. Monitoring copyright involves the generation and analysis of data regarding the amount and frequency of creation, dissemination, and use of copyrighted materials. Controlling copyright involves using this information through some mechanism acceptable in a free society that promotes communication and the increased transfer of information.

7. Any mechanism which is set up to monitor and control the use of copyrighted materials should include among its goals first, the maintenance of a "fair return" to copyright owners for the duplication of multiplication of their materials, and second, the maintenance of the greatest amount of public access and use which is compatible with a fair return.

III. METHODOLOGY

1. Introduction

One of the chief outputs of the original Committee to Investigate Problems Affecting Communication in Science and Education (CICP)-- in addition to its efforts in applying the scientific method to research and analyze the copyright situation--was the development of a basic two-dimensional matrix displaying vertically, various proposed solutions to the copyright/photocopying problem, and horizontally, criteria by which these solutions might be evaluated. The "old" solutions were divided into two groups: "Changes in the Law" and "Administrative Solutions." The old "Changes in the Law" solutions were as follows:

1.) Do Nothing---"Laissex Faire"

2.) Extend the principle of "fair use" to statutory law

3.) Statutory Licensing of Scientific and Educational Copyrighted works: Weak and Strong

4.) Statutory Free Copying

  a.) all Copyrighted works

  b.) Science and Education only

The old "Administrative Solutions" were as follows:

1.) Voluntary Agreement---single copies

2.) Contract Agreement---multiple copies

3.) Non-profit Clearing House

  a.) Statutory

  b.) Voluntary

4.) Ad Hoc Solutions---Publishers

There were twelve criteria used in the original CICP reports, and these were divided into two groups: "Necessary" and "Desirable." The old "Necessary" criteria were the following:

1.) Provides copies promptly

2.) Supports scientific or educational publications

3.) Provides multiple copies

4.) Removes threat of infringement suit

5.) Does not reduce rights to intellectual property

6.) All interested groups benefit in solution

7.) Not contrary to Universal Copyright Convention (UCC)

8.) Requires no distinction between science and educational works and other copyrighted works

The old "Desirable" criteria were as follows:

1.) Short time before starting

2.) Practical to initiate

3.) Practical to administer

4.) Practical to alter

(For a more complete description of the above, see "First Annual Report of the Committee to Investigate Problems Affecting Communication in Science and Education" in Bulletin of the Copyright Society of the U.S.A., vol. 1, no. 1, Oct. 1962, pp. 1-25.)

Those solutions which satisfied a specific criterion were assigned a "1"; those solutions which did not satisfy a specific criterion were assigned a "0". Each criterion was applied against each solution, and at the end, the CICP added up the "ones" across rows to develop a rough estimate of the ranking of each solution. Each rating was examined by several experts in the field of copyright, and a consensus reached. According to this original, rudimentary "decision table," the Non-profit Clearing House satisfied the most criteria, and was described in somewhat more detail than the other solutions.

As mentioned in the introduction, many changes have occured since the CICP reports. With recurring pressure for Congressional legislation in the

94th Congress, it seemed an opportune time to reassess the work of the CICP and press on. Therefore, Professor Laurence B. Heilprin of the University of Maryland, the original chairman of the CICP study group, assembled a team of graduate students to attack the copyright/photocopying problem.

2. Phase I: Formulation of Objectives and Review of Relevant Literature

Initially, the project goals were outlined as:

1.) Review CICP report; compile a bibliography of relevant material provided through the courtesy of the copyright office

2.) Update or replace original CICP solutions and criteria

3.) Develop a ranking scheme, if possible, to weight individual criteria

4.) Estimate respective ranking of suggested solutions

5.) Describe organizational/legislative requirements for "best" solutions

6.) Outline benefits of top-ranking solutions and make suggestions for implementation.

Tasks (1), (2), and (4) were accomplished. Task (3) was not accomplished because of the potential bias introduced by weighting schemes. There were already two partly subjective factors present. One was the choice of criteria. The other was the assignment of the presence or absence of a criterion in a solution. In assigning the 0's and 1's, if there was any doubt, a question mark was used to indicate this doubt. Tasks (5) and (6) were not completed and remain for another possible project.

With the help of the Copyright Office, we were brought up to date on current developments in copyright, especially as they are affected by photocopying. The catalog and collection of the Copyright Office library was made available to us, and class members searched out literature which dealt specifically with photocopying or proposed solutions, American or foreign.

3. Phase II: Review and Updating of Original CICP Solutions, Criteria and Ranking Scheme

The original CICP criterial distinction between "necessary" and "desirable" was eliminated. Also, the distinction between scientific and educational publication and other publication was dropped. However, the distinction between for-profit and non-profit use was retained by mentioning non-profit users and non-profit copying agents.

Our final list of criteria was as follows:

1.) Supports creation of new works
2.) Provides copies promptly
3.) Provides multiple copies
4.) Reduces threat of infringement suit
5.) Maintains right of copyright owners
6.) Maintains availability to public-at-large
7.) Consistent with Universal Copyright Convention (UCC)
8.) Does not discriminate against any interested group
9.) Does not violate anti-trust laws
10.) Requires no distinction between profit and non-profit use
11.) Requires no distinction between profit and non-profit copying agents
12.) Short start-up time
13.) Practical to administer
14.) Practical to alter
15.) Infringement easy to control

The solutions were reconsidered in much the same way. Two new solutions were added: "Statutory Option" (suggested by Joan Mavity: see Appendix A) and "Excise Tax on Purchase or Lease of Copying Equipment" (suggested by Benjamin W. Rudd: see Appendix B). Many other newer solutions fell into the "Ad Hoc solutions by Publishers" category. On the whole, however, the original CICP solutions, despite our review of the literature, still seem quite relevant. The main solutions finally agreed upon were as follows (these are defined in Part V):

1.) No change in present law

2.) Fair Use Statutory

    a.) Sec. 107 in S. 22

    b.) Sec. 107 & 108 in S. 22

3.) Statutory Licensing: weak (with "Trigger")

4.) Statutory Licensing: strong (without "Trigger")

5.) Statutory Option

    a.) Type X

    b.) Type Y

6.) Statutory Free Copying: Non-Profit use

7.) Excise Tax on purchase or lease of copying equipment

8.) Non-profit clearinghouse Statutory

9.) Non-profit clearinghouse Voluntary

10.) Voluntary agreement non-profit fair use

11.) Contract agreement--multiple copies on demand

12.) Ad Hoc Solutions by publishers

Methodologically, an important goal in our analysis was to improve upon the original "zero-one" rating for each criterion. In the original report, a solution either satisfied a criterion (one) or it did not (zero). We realized these ratings did not give sufficient attention to the relative importance of the various criteria. However, more important to whom? There were various "parties of interest" involved in the controversy whose opinions would have to be considered:

1.) authors

2.) publishers

3.) technological innovators (e.g., photocopier, telefacsimile, and computer manufacturers)

4.) intermediate disseminators (e.g., librarians, teachers)

5.) users

An attempt was made to evaluate each solution from the point of view of each of the parties involved. This was abandoned, however, because of the great number of possibilities, and because it required value judgments which we felt the reader could make.

Another important aspect was the concept that any proposed solution was a regulatory/control system. The concept of a cybernetic regulatory/control system was used, in part, to assist in development of the assumptions which appear in Part II of this report. A more technical discussion of the copyright control system will be developed separately in the future.

4. Phase III: Evaluation of Outcomes and Reporting

It was soon apparent that the original plan, to sketch the design of the most likely copyright solutions, would have to be left for the future. The process of redesigning the criteria and solutions was sufficiently massive to occupy the Spring term.

IV. DEFINITIONS of CRITERIA

The standards for judging the solutions as used in this report are explained as follows. A number of specifications have been set up which allow for a comparison of the various proposed solutions on an objective basis. The specifications are:

1.) The solution should support the creation of new works in the sciences and the arts, as specified in the Constitution of the United States.

2.) The solution must provide copies promptly: that is, without unnecessary delays or cumbersome clearance procedures. The regulatory system should not retard the capability of the technology to deliver copies.

3.) The solution should provide multiple copies on a routine basis, without requiring separate permissions from copyright owners. Multiple copies are those in excess of fair use, as defined in Senate bill S. 22 of the 94th Congress, 1st session.

4.) The solution should reduce the threat of infringement suit for copying; i.e., it should reduce the threat of litigation in comparison with that under the present law.

5.) The solution should maintain the rights of the copyright owners; i.e., those statutory or judicial rights now enjoyed.

6.) the solution should maintain availability of copyrighted material to the public at large.

7.) The solution should not adversely affect the rights vested in foreign copyright owners entitled to protection under the Universal Copyright Convention (UCC) to which the United States is signatory.

8.) The solution should not discriminate against any interested group. The following groups (neither exhaustive nor mutually exclusive) are recognized as interested parties: authors, publishers, technological innovators, intermediate disseminators, users.

9.) The solution should not violate anti-trust laws. Free competition should be maintained, and trusts, cartels, monopolies, etc., opposed.

10.) The solution should require no distinction between profit and non-profit use.

11.) The solution should require no distinction between profit and non-profit copying agents such as libraries and information networks.

12.) The solution should not require a long delay before it can be put into effect. One critical distinction is whether or not time-consuming legislation would be necessary for initiation. Another cause of delay might be the extent of political and financial support required for initiation. A short start-up time is defined as within three years.

13.) The solution should be practical to administer and economically feasible.

14.) The solution should be practical to alter. Alteration should not require reference to time-consuming legislation or operational reorganization.

15.) The solution should include mechanisms for control of infringement which are easy to initiate and operate.

V. DEFINITIONS of SOLUTIONS

SOLUTION 1: No change in present law if enforced.

Pro:

1. Since this is the status quo, no further effort is required.

2. Satisfactory results are possible for some groups.

Con:

1. This solution does not confront the issue for which a solution is sought. The ever increasing amount of information being produced is accompanied by the need for wider and more immediate dissemination than is possible under the present law.

2. Copyright infringement suit remains a major issue.

3. This solution might hinder the development and utilization of new technology.

SOLUTION 2: Fair Use: Statutory---enact into law the judicial concept of Fair Use. Two types of Fair Use are delineated in the table.

(a) Fair Use: Statutory: Senate Bill S.22, Sec. 107---parts of works may be copied for purposes such as criticism, comment, news reporting, teaching, scholarship or research.

Pro:

1. What has been an indefinite legal concept decided on an individual case by case basis would be more clearly defined and more widely applicable.

Con:

1. Only parts of works may be copied; whole works are not included in this solution.

2. Although this solution may be satisfactory to some groups, many needs extend beyond the partial work allowance stipulated.

3. This section does not address the question of economic

compensation for works copied.

4. The question of how much of a work constitutes a "part" of a work and who would make this decision is left open in this solution.

(b) Fair Use: Statutory: Senate Bill S.22, Sec. 107 & 108---as above, with provisions for library photocopying of the entire work for research and/or archival purposes, provided the library has determined that the work cannot be obtained at a fair price and that a warning of copyright regulations is displayed.

Pro:

1. What has been an indefinite legal concept decided on an individual case by case basis would be more clearly defined and more widely applicable.

2. This solution does not require a distinction between whole works and parts of works.

Con:

1. This section does not address the question of economic compensation for works copied.

2. This section does not provide any practical approach to the problem of multiple and/or systematic photocopying. Since this provision can be interpreted to prohibit inter-institutional loan of photocopies, it may impose severe economic burdens on certain institutions.

3. This solution makes an exception for a particular interest group.

SOLUTION 3: Statutory Licensing: Weak---Copyrighted works cannot be reproduced without permission. However, if permission is given to anyone, others may reproduce the work upon payment of a royalty, set by law, to the copyright owner. In essence, this is a license with a trigger mechanism.

Pro:

1. All forms of copying would be covered in this solution; no differentiation between systematic and non-systematic photocopying would be required.

2. Payment of royalties would be made to copyright owners.

Con:

1. Trigger mechanism introduces two complications: 1) the user would have to determine if the initial permission to copy had been given; 2) those works not licensed could not be photocopied. Foreign works would be excluded from the system.

2. Royalty rates would be difficult to negotiate and/or alter.

3. It would be necessary to create a mechanism for the collection and distribution of royalties.

SOLUTION 4: Statutory Licensing: Strong---any copyrighted work may be reproduced upon payment of a royalty at a legally fixed rate to the copyright owner. Licensing of all works is automatic upon registration.

Pro:

1. All forms of copying would be covered in this solution; no differentiation between systematic and non-systematic photocopying would be required.

2. Payment of royalties would be made to copyright owners.

3. All U.S. works would be licensed and could be copied.

4. No prior determination of initial permission to copy would be necessary.

Con:

1. Royalty rates would be difficult to negotiate and/or alter.

2. It would be necessary to create a mechanism for the collection

and distribution of royalties.

3. Such total control of information dissemination introduces the possibility of abuse.

4. Foreign works would be excluded from this system.

SOLUTION 5: Statutory Option---There would be two types of Copyright: X--Registry with exclusive rights to the copyright material, and Y--Registry with notice of less restricted rights on copying (from one copy to unlimited copies).

Pro:

1. Distinguishes between the two main classes of published materials: those for which the copyright owner desires the work disseminated regardless of renumeration, and works composed primarily for remuneration.

2. Tightens protection for those who want protection.

3. Clarifies the definition of infringement.

Con:

1. Some device for distinguishing the two classes would have to be created.

2. If the author and the registrant are not one and the same, they might differ as to choice of option.

SOLUTION 6: Statutory Free Copying---Permits an unlimited number of copies of copyrighted works for non-profit purposes.

Pro:

1. This solution would not require any additional mechanism for royalty distribution and collection.

2. Certain groups, such as educators, who seek exceptions, would be served by this solution.

Con:

1. This solution does not provide any economic compensation

to the copyright owner for photocopying. Non-profit groups consitute a large portion of the market for copyrighted works. The economic impact of this solution could severely damage the publishing industry. In turn, the economic incentive to create works would be diminished.

SOLUTION 7: Excise Tax on the sale or lease of copying machines, to take the form of a set fee per machine, to be distributed to copyright owners.

Pro:

1. Since payment of this tax authorizes photocopying, the possibility of infringement by photocopying is virtually eliminated.

Con:

1. It is essentially a regressive tax and would discriminate economically against users with low volume copying.

2. Enforcement of this solution would be highly impractical. It would be difficult to control the sale and resale of photoduplicating equipment.

3. The solution would require a mechanism for collection and distribution of revenues.

SOLUTION 8: Non-Profit Clearinghouse: Statutory Membership with anti-trust exemption--A non-profit, privately owned clearinghouse for granting permission on a fee basis to reproduce copyrighted works would be established. Upon registration for copyright, all works would be included in the system.

Pro:

1. All copyrighted works would be included in the system.

2. The system would allow for systematic reproduction and distribution of copyrighted material.

3. Economic compensation would be made to copyright owners.

Con:

1. A mechanism for collection and distribution of royalties must be created.

2. Such extensive centralization of information dissemination could disrupt the free flow of information.

SOLUTION 9: Non-Profit Clearinghouse: Voluntary Membership with anti-trust exemption---A non-profit, privately owned clearinghouse for granting permisiion on a fee basis to reproduce copyrighted works would be established. Copyright owners would participate voluntarily.

Pro:

1. Economic compensation would be made to copyright owners.

2. The solution would allow for the systematic reproduction and distribution of copyrighted material.

3. The fact that the system is voluntary retains the discretionary rights of the copyright owner.

4. Voluntary agreements allow greater flexibility in setting royalty rates.

5. If a few influential publishers join, there will be a bandwagon effect on membership.

Con:

1. All works are not necessarily included in the system. In order for this system to be effective most copyright owners would have to participate.

2. Prior determination is necessary as to whether the material to be copied is in the clearinghouse.

SOLUTION 10: Voluntary Agreement: Non-profit Fair Use---Agreement among copiers and copyright owners to permit copying within the Fair

Use guidelines of S.22.

Pro:

1. Since this solution is voluntary it permits a greater flexibility for negotiation between interested parties.

2. No additional mechanism for the collection and distribution of royalties is necessary beyond that which the parties to the agreement provide.

Con:

1. All copyrighted works are not included in this system.

2. The user must determine, before copying, whether the material is covered by a voluntary agreement.

3. This solution does not provide any practical approach to the problem of multiple and/or systematic photocopying. Since this provision can be interpreted to prohibit inter-institutional loan of photocopies, it may impose severe economic burdens on certain institutions.

SOLUTION 11: Contract Agreement: On-demand multiple copies---Agreement on a contract basis, between copyright owners and intermediate disseminators, such as libraries, to permit a limited number of copies, provided payment of royalties is made.

Pro:

1. The system would allow for systematic reproduction and distribution of copyrighted material.

2. Economic compensation would be made to copyright owners.

3. The mechanism for distribution of royalties is handled by the copyright owner and/or the agent.

4. Contractual basis provides for adaptation to changing technology.

Con:

1. All copyrighted works are not necessarily included in this

system.

2. Individually negotiated contracts may be more costly in the aggregate than a uniform solution.

SOLUTION 12: Ad Hoc Solutions by Publishers---two types of agreements are included.

(a) Agreements between publisher and user in which the publisher supplies photocopies and/or reprints of copyrighted works. Generally, these would be copies provided at the request of the user on a fee basis.

Pro:

1. Publisher and/or copyright owners retain exclusive control of their works.

2. Economic compensation is made to the copyright owner.

3. No distinction between systematic and non-systematic photocopying is needed. Individual agreements could be reached for different types of users.

Con:

1. It is unrealistic to expect present photocopying practices in libraries and other institutions to be abandoned.

2. Exclusive control of information access by these means could disrupt the free flow of information.

(b) A two-tier system of subscription payment, one price for individual use and another, higher price for institutional subscribers which, in effect, includes permission to copy.

Pro:

1. Economic compensation is made to copyright owners.

2. This solution eliminates the necessity for extra mechanisms to collect and distribute royalties.

3. The system would allow systematic reproduction and distribution

of copyrighted material.

4. Since this payment is collected and distributed to copyright owenrs before copying occurs, the possibility of infringement by photocopying is reduced.

Con:

1. The uses that the higher price would allow would not be clearly defined and might be restrictive.

2. All works are not in the system; only those with a subscription price would be included.

3. This solution might become a precedent for other restrictions; e.g., library and network circulation.

VI. <u>TABLE</u>: Zero-One Analysis of Proposed Solutions to Copyrighted Photocopying Problem by Selected Criteria

SOLUTIONS *	SUPPORTS CREATION OF NEW WORKS	PROVIDES COPIES PROMPTLY	PROVIDES MULTIPLE COPIES	REDUCES THREAT OF INFRINGEMENT SUIT	MAINTAINS RIGHTS OF COPYRIGHT OWNERS	MAINTAINS AVAILABILITY TO PUBLIC-AT-LARGE	CONSISTANT WITH UCC	DOES NOT DISCRIMINATE AGAINST ANY INTERESTED GROUP	DOES NOT VIOLATE ANTI-TRUST LAWS	REQUIRES NO DISTINCTION BETWEEN PROFIT AND NON-PROFIT USE	REQUIRES NO DISTINCTION BETWEEN PROFIT & NON-PROFIT COPYING AGENTS	SHORT START-UP TIME	PRACTICAL TO ADMINISTER	PRACTICAL TO ALTER	INFRINGEMENT EASY TO CONTROL
NO CHANGE IN PRESENT LAW	1	0	0?	0	1	0	1	0	1	1	1	1	0	0	0
FAIR USE: SEC. 107 in S. 22	1	1	0	1	0	0	0	0	1	0	0	0	1	0	0
STATUTORY:SEC. 107 & 108 in S.22	0	1	0	1	0	1	0	0	1	0	0	0	1	0	0
STATUTORY LICENSING: WEAK (with "TRIGGER")	0	0?	1	1	0	1	0?	0	1	1	1	0	1?	0	0
STATUTORY LICENSING: STRONG (without "TRIGGER")	0	1	1	1	0	1	0	0	1	1	1	0	1?	0	1?
STATUTORY OPTION X	1	0	0?	0	1	0	1	0	1	1	1	0	0	0	1?
STATUTORY OPTION Y	1	1	1	1	1	1	1	1	1	1	1	0	1	0	1?
STATUTORY FREE COPYING: NON-PROFIT USE	0	1	1	1	0	1	0	0	1	0	0	0	0	0	0
EXCISE TAX ON PURCHASE OR LEASE OF COPYING EQUIPMENT	0	1	1	1	0	1	0	0	0?	1	1	0	1	0?	1
NON-PROFIT CLEARINGHOUSE STATUTORY MEMBERSHIP	0	1	1	1	0	1	0	0	1?	1	1	0	1	0	1
NON-PROFIT CLEARINGHOUSE VOLUNTARY MEMBERSHIP	1	1	1	1	1	1	1	1	0?	1	1	1	1	1	0
VOLUNTARY AGREEMENT NON-PROFIT FAIR USE	1	1	0	1	1	1?	1	1?	1	0	0	1	1	1	0
CONTRACT AGREEMENT MULTIPLE COPIES ON DEMAND	1	1	1	1	1	0	1	1?	0?	1	1	1	1	1	0
AD HOC SOLUTIONS BY PUBLISHERS	1	0	1	1	1	1?	1	0?	0?	1	1	1	1	1	0

*1: CONDITIONS SUBSTANTIALLY MET 0: CONDITIONS NOT SUBSTANTIALLY MET ?: DOUBTFUL

VII. CONCLUSION

1. Solutions

The table shows that three solutions scored highest among the twelve options. The Statutory Option is neither weak nor strong; however, it clearly indicates what may be copied. Without modification, however, it may be unacceptable to publishing interests. The Contract Agreement is the easiest of the three to implement, and seems widely acceptable. The Non-Profit, Voluntary Clearing House stands out because it centralizes royalty collection and distribution for materials, under direction of a neutral administration. It could apply to other classes of copyright and also appears widely acceptable.

2. Unsolved Problems

1) Not yet fully understood:

a) The relationship between copyright protection and the stimulus it provides authorship or dissemination in the sciences or arts.

b) Problems incurred by conversion from one medium to another; e.g., input into computer systems and networks.

c) Social, political, and psychological reactions to changes in information transfer as affected by copyright.

d) Effect of copyright on growth of library and information networks.

2) Not sufficiently substantiated nor defined:

a) Financial damage attributed to photocopying.

b) "Systematic Reproduction" as referred to in Senate Bill S.22, 94th Congress, 1st session.

3) Congressional action may be required:

a) Further amplification of the definition of fair use of copyrighted works (see 2b, above).

b) Legislation for a possible Statutory Option solution

(see Appendix A) or an Excise Tax (see Appendix B).

c) Charters and/or anti-trust exemptions for possible solutions by Contract Agreement or by Clearinghouse.

d) Statutory accommodation for future changes in design and use of reprographic, computer, and other communication technology.

3. Future Directions

Future progress will depend on how objectively, scientifically, and cooperatively the unsolved problems---listed and unlisted---are approached. The problems must be recognized as increasing in magnitude more rapidly than those to which traditional legislative methods can respond.

APPENDICES

APPENDIX A

STATUTORY OPTION

by Joan Mavity

The 1909 Copyright Law is inappropriate because it gives the same rights to two different types of writing that have different goals. The law should be rewritten in a manner that will support both types of authors and which, at the same time, will protect the interests of the copyright owner and the general public.

This paper proposes a solution to the photocopying-copyright dilemma and states the rationale therefor. The proposed solution is to establish two basic types of copyright, the types being distinguished by the degree of exclusiveness retained by the owner in the notice of copyright. In one type, the owner would retain all rights without exception. In the other, the owner would grant rights to copy to an extent indicated in the notice. This statement could relinquish all rights to copy; or grant permission to make single copies for non-profit purposes; or delineate any degree between the two extremes.

A major cause of the photocopying-copyright dilemma is the failure of the law to distinguish between two main types of published materials. In a literary work, the form and substance are indissoluble; when the form of such a writing is changed, the work is destroyed. Paraphrasing a section from a novel, essay or poem changes the form and not the substance, but in effect, destroys the work, because the form is as essential to the integrity of the section as is the substance.

With expository writing, however, the substance constitutes the creation, and the form is merely a means of communicating the substance. Paraphrasing a paragraph from an article, describing a new laboratory method, does not destroy the paragraph because the integrity of the paragraph lies in the substance of the information it conveys rather than in the form by which the information is expressed.

The author of a literary work writes to support himself and therefore is concerned that his writing be sold and that it not be freely copied. The expository writer is not dependent upon the sale of his articles for his income; he is a researcher or scholar who wishes to communicate his discoveries or inventions to others in his field. He wishes to make known his contribution to his field and the publication of his articles is the method by which he disseminates his ideas.

Whereas some authors wish to be paid when their writings are copied, others not only do not wish to be paid, but actually purchase copies of their own works in order to give them away. Few authors own the copyright on their work; they relinquish the right in order to be published. The copyright owners, who bear the expense of putting the works into print, retain the exclusive right to copy, in order to recover the costs of publishing and in order to make a profit. This state of affairs has led to a situation in which both the literary writer and the writer of expository prose are afforded the same rights, in spite of the fact that their goals are completely different.

The fact that the copyright law and regulations do not allow for the different motivations of the two types of writers, has led to a breakdown of the ideal inherent in the conception of copyright, which is to promote the progress of writing. The increasing sophistication of reprography technology, along with the need of the general public for unrestricted and immediate access to certain copyrighted works, reinforces the seriousness of the breakdown of the copyright system; and such sophisticated advances heighten the necessity of changing the established method of dealing with the copying of copyrighted material. No segment of the concerned public--neither authors, publishers, librarians, educators, nor the users--wishes to subvert the system provided for in the Constitution. Today the courts realize that researchers need to photocopy journals; and yet, the leniency of the courts in this respect compromises the interests of writers of literary works who do not want their writing freely copied and distributed to people who do not buy their books. Researchers support the publishers who accept their articles so that information in the articles may be disseminated; but the expository writer's publisher forbids the public to copy their client's work because such copying decreases the publisher's profit.

The system must be changed so that the author can choose the type of copyright he needs to protect his work. There must be a method established which allows works whose usefulness depends upon widespread dissemination to be copied without penalty, and at the same time, such a method must give clear protection to published works from which the author's income is derived.

The solution proposed here relies on the premise that copyrighted expository writing is more frequently copied than are literary works, and that it is expository prose that the general public--the users--need immediately on demand. The law should be flexible enough to permit expository writing to be copied without infringing upon the rights of the writer of literary works.

When a work is published there should appear, in the notice of copyright, a statement of the exclusive right retained by the owner. The statement should explicitly limit the extent to which the publisher and author will permit the public to copy the work. If the notice retains the right to copy, the copyright law should be enforced without such exceptions as "fair use," "gentlemen's agreement," and similar attenuations; only the copyright owner should have the authority to grant users permission to copy. The law should define the variations possible in the copyright notice and should state what constitutes infringement in a clear manner so that it will not be necessary to make attenuating interpretations of the law.

With few exceptions, publishers of literary works will probably retain the right to copy. Publishers of expository writings should grant the right to copy if the thesis is correct and they should compensate for the loss of income they incur in permitting the material to be copied by increasing their charges to the institutions which make the copies.

If the law delineates the variations permissible within the copyright notice, and if the law states what constitutes infringement in a clear manner, then the purpose of copyright will be fulfilled by protecting both types of authors from illegal copying; and the general public will be able to use the writings of all authors, without breaking the law.

REFERENCES

1. Heilprin, Laurence B., Alternate solutions. In Hattery, Lowell H. and Bush, George P. (eds.), Reprography and copyright law, Washington, D. C., A.I.B.S., 1964.

2. Heilprin, Laurence B., Response to request for comments on the proposed bill to establish a National Commission for Study and Recommendations on Copyright, unpublished manuscript, University of Maryland, July 10, 1967.

3. Wellek, Rene and Warren, Austin, Theory of Literature, New Revised Edition, Chapter 3, Harcourt, Brace, 1956.

APPENDIX B

EXCISE TAX SOLUTION

As an alternative to the photocopying problem, it is proposed to levy:[1]

> "...a Federal excise tax on sales of all apparatus capable of reproducing copyrighted materials, the proceeds to be used for the ultimate distribution to creators and performing artists through the media of already existing author and performing organizations.
>
> The validity of such a tax could be justified under the constitutional power of Congress to levy taxes for the general welfare.[145] And the encouragement of the creative and performing arts is in the public interest.[146]
>
> It has been stated that the power of Congress to impose excise taxes is subject only to the limitation that they be for the public welfare and uniform throughout the United States.[147]

---

[1]Benjamin W. Rudd, General Attorney-Librarian, Copyright Librarian, Copyright Office, Home Recording Devices and Copyright, Unpublished manuscript, June 1965, pp. 55-56.

[145]See United States vs. Butler, 297 U.S.1, 65(1936); Helvering vs. Davis, 301 U.S.619, 640 (1937).

[146]"The economic philosophy behind the clause empowering Congress to grant patents and copyrights is the conviction that encouragement of individual effort by personal gain is the best way to advance public welfare through the talents of authors and inventors in 'Science and Useful Arts.'" (emphasis supplied) Mazer vs. Stein, 347 U.S.201, 219 (1954).

[147]See, e.g., Flint vs. Stone Tracy Co., 220 U.S. 107, 153 (1911); Chas. C. Steward Mach. Co. vs. Davis, 89 F2d 207, 209 (5th Cir.), aff'd, 201 U.S. 4 548 (1937).

REFERENCES

SOLUTION 1: No change in present law

1. "Committee to Investigate Copyright Problems Affecting Communication in Science & Education. First Annual Report, (CICP) May 10, 1960." Bulletin of the Copyright Society of the U.S.A., vol. 10, no.1, October 1962.

SOLUTION 2: Fair Use ,Statutory

2. Bender, Ivan, "Copyright: Chaos or Compromise?". Previews, November 1973, p. 3-5.

3. Bender, Ivan and David Engler, "When is it Legal to Duplicate?". Media and Methods, January 1975.

4. Harvell, Michael C., "Copyright--Fair Use--Government Photocopying of medical journals does not infringe Journal Copyright." Boston University Law Review, May 1974, pp. 689-701.

5. Henry, Nicholas L., "Copyright: Its Adequacy in Technological Societies." Science 186:993-1004, December 13, 1974.

6. Marke, Julius J., "1962 Report of Representative to the Joint Libraries Committee on Fair Use in Photocopying." Law Library Journal 55:185, May 1962.

7. Marke, Julius J. International Federation of Library Association/ International Association of Law Libraries Working Group On Copyright, unpublished report, 1974.

8. Maslansky, Harris, "Copyright: What Copyright is Right?" Science 40: 49-50, May 1973.

9. Passano, William, "A Publisher's View of Photocopying." American Libraries, May 1974, pp. 197-208.

10. U.S. Senate. Senate Bill S.22, January 1975. 94th Congress, 1st Session.

11. Willig, Peter F., "Reprography and the Law." Reproductions Review 15:23-30, February 1965.

SOLUTION 3: Statutory Licensing--Weak

12. op.cit., CICP Report. (Solution 1)

SOLUTION 4: Statutory Licensing--Strong

13. Authors League of America, Inc., "A Licensing System." Library Journal February 15, 1966, pp. 892-893.

14. Committee on Scientific and Technical Information. The Copyright Law as it Relates to National Information Systems and National Programs, April 1967.

15. Karn, Irwin, "A Statutory Licensing System for the Limited Copying of Copyrighted Works." Bulletin of the Copyright Society of the U.S.A. 12:197-209. 1965.

16. op.cit., IFLA/IALL Working Group on Copyright. (Solution 2)

17. Nimmer, Melville B., "New Technology and the Law of Copyright: Reprography and Computers." UCLA Law Review v. 15, 1968, pp. 931-975.

18. "Photocopying and Fair Dealing," The Bookseller 3585: 1728-1732, September 7, 1974.

SOLUTION 5: Statutory Option

19. Mavity, Joan. "Statutory Option" (see attached Appendix A)

SOLUTION 6: Statutory Free Copying--Non-Profit Use

20. op.cit., CICP Report. (Solution 1).

SOLUTION 7: Excise Tax on Purchase or Lease of Copying Equipment

21. Mixed Study Committee Photomechanical Reproduction, "About Photomechanical Reproduction." Interauteurs 175-179, 1969-1970, pp. 121-134.

22. op.cit., Melville Nimmer. (Solution 4).

23. Rudd, Benjamin W., "Home Recording Devices and Copyright: unpublished paper, June 1965.

SOLUTION 8: Non-Profit Clearinghouse, Statutory

24. op.cit., CICP Report. (Solution 1).

SOLUTION 9: Non-Profit Clearinghouse, Voluntary

25. op.cit., Committee on Scientific and Technical Information. (Solution 4)

26. op.cit., CICP Report. (Solution 1).

27. Heilprin, Laurence B., "An Experimental Copyright Moratorium: Study of a Proposed solution to the Copyright Photocopying Problem." A final report to the American Society for Testing and Materials (ASTM) by the Committee to Investigate Copyright Problems Affecting Communication in Science and Education (CICP).

28. Heilprin, Laurence B., "Technology and the Future of the Copyright Principle." Phi Delta Kappan, January 1967, pp. 220-225. Am. Doc., Jan.1968, 6-11.

29. op.cit., Mixed Study Committee Photomechanical Reproduction. (Solution 7).

SOLUTION 10: Voluntary Agreement,Non-Profit Fair Use

30. op.cit., CICP Report. (Solution 1).

SOLUTION 11: Contract Agreement--multiple copies on demand

31. Berne Union "Working Group on Reprographic and Reproduction of Works Protected by Copyright---Report," Copyright, September 1, 1973.

32. op.cit., CICP Report. (Solution 1).

33. Gehlin, Jan. The Swedish Writer and His Rights. Swedish Institute, Stockholm. 1973.

34. International Business Machines Corp., "License Agreement for IBM Program Products."

35. op.cit., Mixed Study Committee Photomechanical Reproduction. (Solution 7).

36. "Photocopying: How is Europe Handling the Problem?" Publishers Weekly, November 25, 1974., pp. 26-27.

37. Swedish Literary Commission. Summary of the Main Report Stockholm, 1974.

SOLUTION 12: Ad Hoc Solutions by Publishers

a) Agreements between publisher and user

38. Heilprin, Laurence B., "The Economics of 'On Demand' Library Copying." National Microfilm Association Proceedings, 1962, pp. 311-339.

39. Heilprin, Laurence B., "Technology and the Future of the Copyright Principle." Phi Delta Kappan, January, 1967. (Solution 9).

b) A two-tier system of subscription payment

40. Day, Melvin S., unpublished remarks on Copyright and Library Photocopying given at ASIS Student Chapter Meeting, Catholic University, March 20, 1975.

41. op.cit., Mixed Study Committee Photomechanical Reproduction. (Solution 7).

GENERAL SELECTED BIBLIOGRAPHY

Breyer, Stephen, "The Uneasy Case for Copyright: A Study of Copyright in Books, Photocopies, and Computer Programs." Harvard Law Review 84:281-361, December 1970.

Bush, George P., ed., Technology and Copyright: Annotated Bibliography and Source Materials. Mt. Airy, Maryland: LoMond Systems, Inc., 1972.

Cambridge Research Institute, Omnibus Copyright Revision: Comparative Analysis of the Issues. Washington, D. C.: American Society for Information Science (ASIS), 1973.

Hamilton, Clark. February 11, 1975 informal talk on Copyright, Cable television and Computer Networks deliverd at University of Maryland for LBSC 858-H. unpublished (tape available at the College of Library and Information Services at University of Maryland, College Park)

Hattery, Lowell H. and Bush, George P., eds., Reprography and Copyright Law. Washington, D. C.: American Institute of Biological Sciences, 1964.

Heilprin, Laurence B., "Response to Request for Comments on the Proposed Bill to establish a National Commission for study and Recommendations on Copyright." unpublished paper, University of Maryland, July 1967.

Ringer, Barbara A. January 24, 1975, informal talk on history of Copyright, delivered for University of Maryland LBSC 858-H class. unpublished. (tape available at College of Library and Information Services at University of Maryland)

Ringer, Barbara A. February 7, 1975, informal talk on where Copyright stands today, delivered for LBSC 858-H class. unpublished. (tape available at College of Library and Information Services at University of Maryland)

COPYRIGHT CLEARINGHOUSE:
IN-DEPTH STUDY OF A PROPOSED SOLUTION TO THE COPYRIGHT PHOTOCOPYING PROBLEM

By the class in Operations Research for
Library and Information Services
(LBSC 858 II)

College of Library and Information Services
University of Maryland
College Park, Maryland

Spring 1976

Graduate Students

Earl J. Dotterweich
Thomas L. Eichman
Nancy K. Holden
Peri L. Schuyler
Hyon J. Woo

Faculty

Laurence B. Heilprin

TABLE OF CONTENTS

ABSTRACT

A solution to the copyright photocopying problem is offered in the form of a private, non-profit, voluntary copyright clearinghouse with equal participation in its direction by publishers and users. A model is designed for exchange of permissions to copy for royalty fees based on numbers of copies made. A system of statistical sampling is outlined to insure fair distribution and to greatly reduce recordkeeping overhead. The program calls for two stages: organization of the clearinghouse (start-up) and stable operation (steady state). The latter is economically sensitive to volume of copying. Under conservative assumptions the projected U. S. annual volume of copied copyrighted pages will reach the specified steady state support level by 1980. It is concluded that organization of a copyright clearinghouse, mutually advantageous and self-maintaining, is feasible now.

The versatility of such a system for interchange of fees and permissions points to a third possible stage, extension to copyrighted material using non-print media and other means of communication emerging from new technology.

COPYRIGHT CLEARINGHOUSE:

IN-DEPTH STUDY OF A PROPOSED SOLUTION TO THE COPYRIGHT PHOTOCOPYING PROBLEM

I. INTRODUCTION

This study is the product of a graduate class in Operations Research conducted in the College of Library and Information Services at the University of Maryland during the Spring semester of the 1975-1976 school year.

The subject is the conflict between the holders of copyrighted material and those who gain access to this material through photoduplication. At the center of the controversy is the revision of the Copyright Law now going on in Congress, and its impact on both sides of the issue [1] [2].

In the last several years, the literature has been rife with elaborations of the "photocopying problem". Eloquent defenses of the author-publisher viewpoint are equally countered by those of the user (library, research institution, educational institution), and, indeed, there are merits to the arguments of each [3] [4]. Another significant portion of the literature has been given over to suits for infringement, challenges to current law, court decisions, and opinions on these decisions.

Two things become evident from this state of affairs: 1) Along with the plethora of literature dealing with the problem, there is a dearth of literature dealing with the solutions to the problem; and 2) the Copyright Law, like many other laws, is open to interpretation, and as long as there is room for interpretation, there will be conflicts, apparently doomed to court settlements on the merits of each individual case.

Last year's class in Operations Research prepared a survey of the proposed solutions to the copyright controversy, setting out the advantages and disadvantages of each and weighting each according to how it met requirements for free flow of information, as well as fair return to the creators of information. With this report (Proposed Solutions to the Problem of Photocopying Copyrighted Material: a Survey) [5] as well as two earlier papers, the original report of the Committee to Investigate Copyright Problems Affecting Communication in Science and Education (CICP) from May, 1960 [6], and a 1963 working paper on clearinghouse feasibility [7] and a 1970 paper on actual trial [8], serving as frames of reference, this year's class undertook an in-depth analysis of the Copyright Clearinghouse (CHC), one of a few solutions, according to last year's class, which met a maximum number of requirements.

The class wishes to express its warm appreciation for generous contributions in time, lectures, and guidance to the following people: Wilma S. Davis, Attorney-Advisor-Librarian, U.S. Copyright Office; Waldo H. Moore, Chief, Copyright Reference Division, U.S. Copyright Office; Michael S. Keplinger, Senior Attorney for the National Commission on New Technological Uses of Copyrighted Works (CONTU); Dr. Harold E. Bamford, Jr., Program Director, Access Improvement Program, Division of Science Information, National Science Foundation; and Donald W. King, Vice President, Market Facts, Inc. Thanks are also given Ms. Joan P. Mavity, U.S. Patent Office, for literature searches; and to Ms. Eleanor B. Waters, Office Supervisor, Department of Computer Science, University of Maryland, for typing and reproducing the manuscript.

II. ASSUMPTIONS

The following assumptions underlie the problem:

1) Provision of a mechanism fostering the unhindered flow of information should be the chief goal of creators and users of information alike.

2) The present economic situation and the (unlimited) opportunity for interpretation of the Copyright Law combine to make publishers and users at once intractable opponents and inescapable partners in the struggle for economic survival.

3) Establishment of an independent clearinghouse as a voluntary, non-profit association of the two main elements in society involved represents an attempt to define a workable solution to the photocopying problem.

4) Since the doctrine of "fair use" as described in the proposed revision of the Copyright Bill (S. 22, Sections 107 and 108) still leaves open to judicial interpretation the degree to which photocopying of copyrighted material is to be allowed, the clearinghouse should refrain from any attempt to determine what proportion of any user's or user institution's copying falls within the domain of "fair use"; unless there is an a priori agreement on the part of publishers allowing a certain percentage of all copied material ascribed to them to fall within the scope of "fair use", and thus exempt from royalty fees (Section III A-1).

5) Following this reasoning there would be no attempt to categorize the type of photocopying done, (save distinguishing between copyrighted and non-copyrighted material). "Single copies" and "systematic reproduction" alike would be tabulated in clearinghouse calculations of fee assessment.

6) Full recording, and therefore policing, of all photocopying transactions is neither desirable nor economically feasible, and would not be attempted. Rather, systematic sampling studies carried out at user institutions, and voluntary data submission by publishing houses, would form the basis of calculations for fee collection and royalty disbursement.

7) Copying-member institutions of the clearinghouse would agree to allow sampling operations to be carried out at any time without notice; the

clearinghouse would conduct these samplings with minimum disturbance to work schedules. This procedure would allow the clearinghouse to sample transactions at random, with least inaccuracy in measurements of distribution.

8) Reproduction of copyrighted material by established demand publishers, e.g., scholarly reprint houses, would not ordinarily concern the clearinghouse. These firms have already established mechanisms for acquiring rights to copyrighted material [9].

9) The clearinghouse would not serve as agent between publishers and authors of copyrighted works. The process of author-publisher royalty negotiation would continue in the traditional practice of individual bargaining.

10) The organization of the clearinghouse would offer compelling inducements for all parties concerned: producers of copyrighted materials, users of the materials, the public. Each would be represented in a balanced governing board. For producers, annual return would be equitably distributed, and high enough to be a strong incentive to join. For users, multiple copies could be made without fear of infringement, while payments would be simplified through one uniform fee per page copied; and the clearinghouse would have the right and duty to reduce the copying fee when economics permitted. The public would be served through a new, flexible channel of communication designed to enhance "the Progress of Science and the useful Arts".

11) Since the most critical parameter determining clearinghouse feasibility is total national volume of copyrighted works copied per year (Section V), and this volume, while increasing, is only entering the lower feasibility threshold, successful establishment of a clearinghouse operating on a national scale would require initial protection, i.e., Congressional exemption for a suitable period from liability to prosecution under antitrust law.

III. ORGANIZATION OF THE CLEARINGHOUSE

The above assumptions while neither exhaustive nor mutually exclusive, underlie development of a clearinghouse designed solely to act as a central agency of exchange between the creators and the dispensers of copyrighted works and the users of such works. Since information as a resource is not independent, but interdependent on both the ability to create and disseminate knowledge and the ability to acquire and retrieve knowledge, the objectives of the clearinghouse are to establish and maintain the flow of information equitably and profitably to creator and consumer alike.

In order to perform its duties as efficiently as possible, it is envisaged that the clearinghouse be established as a private corporation with a Congressional charter, with private capital formation (possibly assisted by a Government loan repayable after reaching the self-maintaining state). The clearinghouse would act as a non-profit middleman, acquiring photocopying rights from publishers, and dispensing these rights to user institutions such as libraries, educational and research facilities. It would collect from these users royalty payments based on established schedules, which in turn it would distribute back to the publishers of copyrighted materials. In other words the clearinghouse would, at least to start, act as a clearinghouse between organizations (publishers and users) rather than individuals -- both kinds of organization dealing regularly ("systematically") with large numbers of copyrighted materials as <u>agents</u> for the creating or the using individuals. Direct contractual relation with individual users would be a second step beyond the present systems design, but dependent on a successful first step. The following represents one, not the only, approach to a feasible solution.

A. Administration of Clearinghouse

Policy for administration of the clearinghouse would be a set by a governing board. The board would be comprised of an equal number of representatives from the publishing community and from the user community. Members would be elected for finite terms. The board would have an ex officio chairman, member of a knowledgeable but neutral body -- possibly the Copyright Register or Deputy. Members of the board would be selected mainly from members of the clearinghouse, but by-laws might include representation for individual authors or users.

Through equal numbers of representatives, the board could attain objectivity on policy decisions. Presumably, by accepting a position on the board, a member would come to recognize problems associated with sectors other than his own, and, by subordinating personal demands, effect compromises toward enrichment of the information community as a whole.

The administrative officer, director of the clearinghouse, would report to the governing board. He would have primary responsibility for implementing policy set by the board, in operating the clearinghouse. Personnel reporting to the director would include systems analysts responsible for designing and maintaining smoothly functioning operations, administrative staff, and the associate directors of each of the clearinghouse departments.

In initial planning special care would be taken with the staffing and scope of duties within each department of the clearinghouse. Without concerted interdepartmental communication and exchange of data, an operation on such a potentially large scale might simply grow vast and unmanageable.

A possible outline or organizational hierarchy is shown in Fig. 1. Four main functional groups are suggested: legal, financial, statistical, and computer service departments. Their relations with users and other external groups would (formally) be conducted through the Director.

1) Legal Department

As stated (assumption 3) membership in the clearinghouse would be

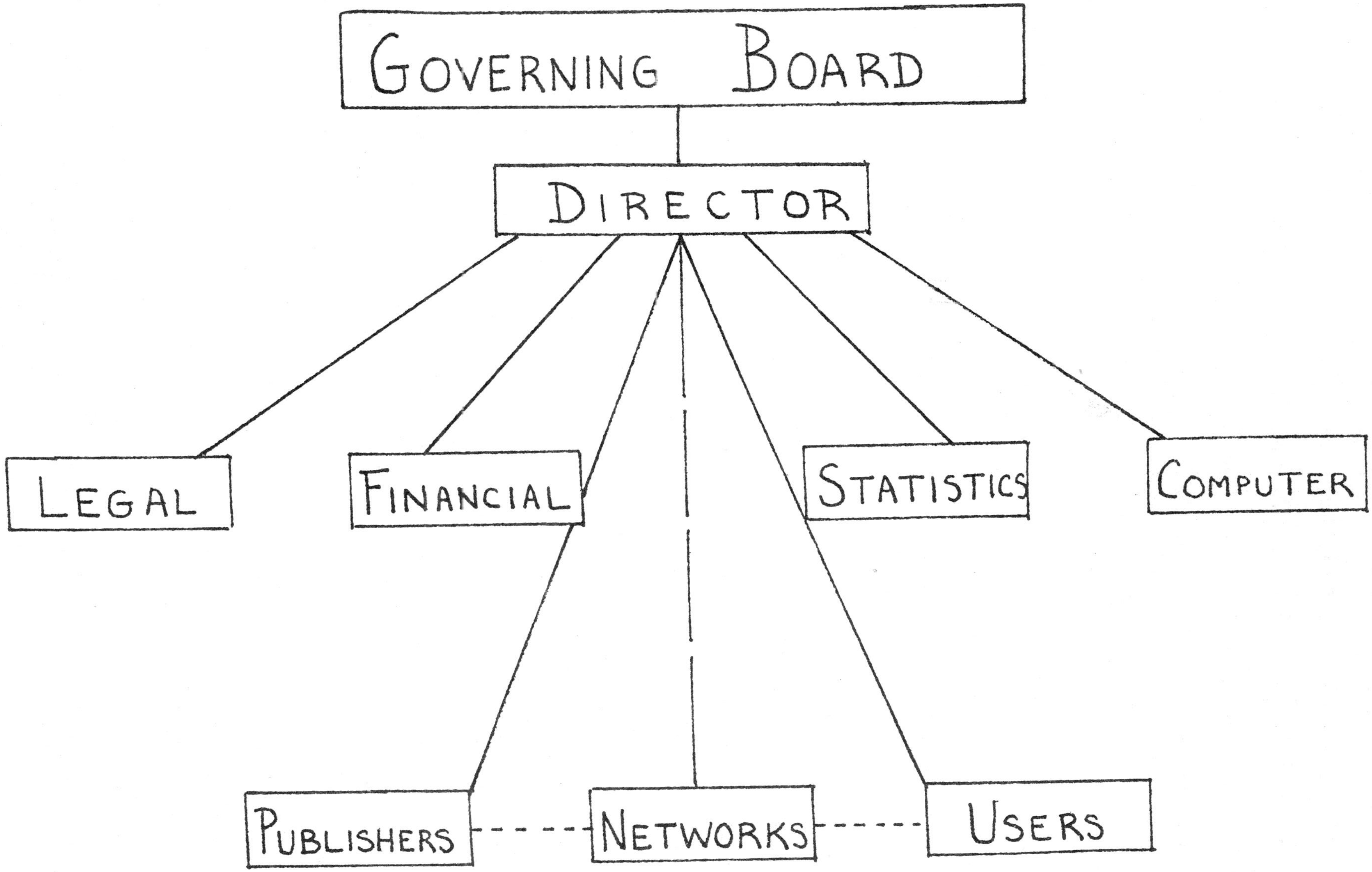
GOVERNING BOARD
DIRECTOR
LEGAL
FINANCIAL
STATISTICS
COMPUTER
PUBLISHERS
NETWORKS
USERS

Figure 1

voluntary, based upon individual contracts between the clearinghouse and consenting publisher, or between the clearinghouse and consenting user institution. The legal staff (with support from the statistical staff) would be responsible for establishing and negotiating the terms of these contracts with the respective participants. Publishers' contracts would be uniform as to method of determining equitable renumeration, including furnishing the clearinghouse with a statistic needed for this purpose (Sec. VB); but might vary as to length of term, liability, etc. Users' contracts would be uniform in granting the clearinghouse permission rights for sampling studies at participating user institutions. It would be understood, defined in the contract, that sampling would be conducted randomly, at unspecified times; that the results so obtained would be confidential and used by the clearinghouse solely for purposes of establishing and maintaining fair assessment schedules for collection or distribution of fees.

It was suggested (assumption 4) that no attempt be made by the clearinghouse to interpret the law or set policy based on those sections of the Copyright Revision dealing with "fair use". Rather, contracts could be drawn with any individual publisher who might wish to allocate a certain proportion of all copied material from his publishing house to "fair use", and thereby exempt that portion from royalty fee. This aspect would be open to re-negotiation at the end of any contract term, and could be added to or deleted from any prior contract.

Other suggested provisions in contract terms between publishers and the clearinghouse are: permission to assign rights to copy in whole or in part, of any output from the publishing house; clearinghouse setting of uniform per page royalty rates; form and frequency of royalty payments.

For contracts between user institutions and the clearinghouse, suggested provisions are: in consideration of compliance with contract terms, guaranteed protection from infringement suits by participating publishers; unlimited rights to copy regardless of number of copies or portion of a whole work to be copied; form and frequency of payments to the clearinghouse.

Both the legal and administrative staffs would have responsibility for maintaining liaison with the Congress, the Copyright Office, the National Commission on New Technological Uses of Copyrighted Works (CONTU), the National Commission for Libraries and Information Services (NCLIS), and other agencies in order to keep the clearinghouse abreast of the current status of copyright law, and to keep informed those members of the information community at the national level on the progress of the clearinghouse.

The legal staff would advise the director on matters of law, and if necessary, represent the clearinghouse at law. The staff would also negotiate with existing library and computer networks to establish procedures for mutual access. This aspect will be further discussed in Section A(4)(a).

2) Finance Department

The operations of the finance department would fall into two broad categories: internal - budgetary and payroll obligations for the clearinghouse; and external - billing of user institutions and remission of payments to publishers.

a) INTERNAL: This section of the department would be staffed with budget analysts who would work with the office of the director of the clearinghouse to establish manpower, equipment and budgetary needs within the framework of the policy established by the governing board. As initial costs will be higher than the costs of maintaining established protocols, this section will also be responsible for projecting future changes and adjusting manpower and budgetary requirements.

b) EXTERNAL: The staff would reduce to a minimum the frequency and effort required by participating members. For example, calling upon the records compiled by the statistics department, threefold routines could be established whereby (1) copying agencies merely report "meter readings" -- total copies made during stated intervals -- perhaps quarterly or semiannually. On this basis and using one, uniform price per copy, they would in reporting also (2) forward to the clearinghouse one check per period. At stated intervals the clearing house would (3) distribute royalty checks to publishers. Thus billing users could almost be eliminated; while information on delinquent accounts could simultaneously be passed to the legal department.

3) Statistics Department

Although no one department would be less vital to the operation of the clearinghouse than any other, successful operation of the whole would depend directly upon the methodologies devised by the Statistics and Computer Services Departments.

Central to the proposed copyright clearinghouse is the concept of counting the number of copies of a given copyrighted work. To carry this out literally, however, is not economcially feasible. It is too costly to keep track of billions of individual transactions, each involving, on the average, only a few cents. The clearinghouse would not pay.

If, however, individual transactions and individual works are not tabulated, but only the publishers of such works (of whom there are less than a thousand); and if the publishers themselves are not individually recorded but sampled, it is possible to maintain an accurate tally at a moderate cost. That is, the main proposal offered here is to replace the concept of individual recording by careful sampling; and the burden of this is borne in an economically feasible manner by the clearinghouse. Sections IV and V discuss the approach.

The statistical staff would be responsible for establishing the pilot procedures for sampling studies; for carrying out the sampling; for conducting studies as required by the clearinghouse. They would tabulate and analyze the data, and from them determine royalty fee assessment and disbursement schedules. Second only to statistical design of the pilot method would be design and execution of periodic update sampling procedures.

4) Computer Services Department

Without efficient computer software capabilities the clearinghouse as a solution to the problem of photocopying would disappear in a jungle of bureaucratic paperwork. In brief outline, the function of the computer department would be to define, establish and implement hardward and software requirements adaptable to all clearinghouse operations. They would include (1) programs for manipulating data derived from statistical sampling studies of copying, and from users and publishers directly; (2) conversion of these data into calculations for fee assessments to user institutions and payment schedules to copyright holders (described in Sec. V); (3) internal operations, such as payroll; and (4) programs for possible linkage to external networks, to extend the services of the clearinghouse or to exchange information.

The computer department would focus the entire operation, acting both as endpoint for acquisition and storage of data and as starting point for processing and dissemination of information. Some of the interactions of the computer department might be as follows:

(a) Computer Services Interaction with Statistics Department: To establish representative samples of user institutions and of publishers. This would require accumulating data on two total populations. The population of user institutions might be broken down by size, location, number of existing publications held, type. The population of publishers might be classed by size, number of existing publications, number of pages produced per year, overall cost of publishing these pages.

Once the total national populations had been stratified, the computer would be used for periodic update sampling calculations that redefined the size and composition of the distribution.

A third stage of computer-statistical operations would be tabulation, storage and use of sampling study results. Algorithms would be designed for projections based on the sample to the total population for setting the copying fee, for payment schedules (weighting the latter in terms of cost of publishing), and for computing the cost of the clearinghouse operation (Sec. V).

(b) Computer-Legal Staff Interactions: Although less extensive than the services provided statistical operations, computer services to the legal staff would still contribute greatly to their operations. Records of contractual agreements would be machine-readable and filed to enable instant recall of contract information. For example, monthly checks could be made on the file to call up contract numbers due to expire, e.g., six months hence, in order to set in motion the process of renegotiation.

(c) Computer-Financial Staff Interactions: Not as complex as statistics requirements, the demands of the financial department would be considerably more varied, although falling into the two broad categories - internal and external operations.

Internal--would include payroll operations, budget, and projections of present needs to determine prospective demands and to forecast patterns of change.

External--through network linkages, public and private, the potential could be explored for expediting accounts information transfer, on-line billing and collecting. Long-term planning could consider other aids in cutting costs of clearinghouse operation and passing them on to copyright owners as larger payments and to user institutions as a lowered fee per page copied.

IV. SAMPLING

As stated, the success of the clearinghouse would depend upon the accuracy and efficiency of the sampling method. The sampling operations would fall into two stages: 1) designing and testing the method and developing a pilot study (initial stage of clearinghouse) based upon those results; and 2) routine carrying-out of the samplings as part of the long-term clearinghouse operations.

The approach falls roughly into the following steps:

1. Definition of objectives of the survey.

To determine, in the case of the user, the number of pages copied per sample period; the number of pages of copyrighted material copied per sample period; and their distributions among publishers. To determine, in the case of the publisher, the total number of pages currently extant, and the average cost/page of publication.

2. Definition of the target populations.

   The total population of users would include libraries, educational and research institutions in both the private and public sectors. The publishing population would include commerical, academic, professional society and other publishers.

3. Establishment of the desired degree of precision with regard to size of the total (target) populations, costs, and time considerations.

4. Determination of the method of measurement.

   As applied to the user population measurement would involve a team of clearinghouse representatives tabulating every piece of material copied at every machine in the institution for the period of the sample (days, weeks). Forms would be devised to account for publisher, number of pages, number of copies, etc. (Fig. 2 and 2a).

5. Stratification of the sample.

   As applied to the user population, breakdown by type (e.g., liraries: academic, public, government, special, etc.), and within each type, by size. Excellent examples of this are found in the Fry-White report [10].

6. Selection of sample size based on the desired degree of precision and the estimated size of the total population.

7. Pretest of the method on a small scale.

8. Summary and analysis of data acquired.

   At this point, the method should be polished, additions or corrections should be made, and the expected amount of residual error defined.

PROPOSED FORM FOR INSTITUTIONAL SAMPLING STUDIES

INSTITUTION: ______________________

DATE: ______________________

MACHINE NO. ______________________

TALLY NO. START: ______ FINISH: ______

NON-COPYRIGHTED	COPYRIGHTED PUBLISHER	TITLE VOLUME/ISSUE/DATE	PAGES	NUMBER COPIES	TOTAL PAGES	BOOK/ SERIAL

Fig. 2

PROPOSED FORM FOR INSTITUTIONAL SAMPLING STUDIES

INSTITUTION: Bicentennial Univ.

DATE: 7/4/76

MACHINE NO. 2

TALLY NO. START: 124 FINISH: 578

NON-COPYRIGHTED	PUBLISHER (COPYRIGHTED)	TITLE VOLUME/ISSUE/DATE	PAGES	NUMBER COPIES	TOTAL PAGES	BOOK/ SERIAL
~~||||~~ ~~||||~~ |	Holden-Day, San Francisco	Steroids, 27/2, 1976	145-53	2	18	s
	Am. Pharmaceut. Assoc., Washington, D.C.	J. Pharmaceutical Sci., 65/3, 1976	339-43	1	6	s
	Blackwell, Oxford & Edinburgh	Immunology for Students of Medicine, 3d ed., 1970	124-37	1	14	B

Fig. 2a

9. Application of results from the sample back to the total population.

The key to the success of the sample is in devising the frames of the sample with respect to size, frequency of sampling and degree of precision such that the projection of sample results back to the total population is within a range of accuracy acceptable to members of the clearinghouse.

V. STEADY-STATE ECONOMICS OF CLEARINGHOUSE SAMPLING AND RATE SETTING

A. Feasibility: As outlined above, in the steady-state the proposed clearinghouse will need several kinds of data for collection and fair distribution of collected fees, and for maintaining its own operations. The main data are: total pages copied from the works of each publisher during the fiscal period, the relative weight to be given each publisher's works (on the basis of cost to produce the pages), and the overhead or budgetary requirement of the clearinghouse. In order to be self-sustaining, the clearinghouse, after paying out nearly all fees collected, takes the remaining fraction of collected revenues, but only within a predetermined (small) value. Definitions and formulas follow. Let

N = the sum of all pages copyrighted by clearinghouse members, and copied by clearinghouse users, per fiscal period (here, one year is assumed). In the model we assume that the total of pages copied in the United States is or is close to the total within clearinghouse contracts. The model, although based on voluntary membership, is designed to attract publishers, so much so that not belonging would not be a trivial disadvantage. Therefore we have not distinguished between N and total national copying. Likewise, we have made no deduction from N for fair use. This is discussed in 2, below.

r_c = the clearinghouse fee charged for copying one page. A feature of the model is that r_c is uniform for all kinds of pages, regardless of actual cost to publish. That is, there will be one price, everywhere the same in the United States in the fiscal period. Object: to simplify fee collection and transfer.

a = fraction of total receipts for clearinghouse maintenance, under steady-state conditions ($0 \leq a \leq 1$). Then,

Nr_c = total receipts of the clearinghouse, in dollars per year (1)

$$Nr_c = N(1 - a)r_c + Nar_c \quad (2)$$

Equation (2) expresses the fractions of total receipts transferred respectively to copyright owners, and to the clearinghouse for its operation. Both the values of r_c and of a are fixed periodically by the clearinghouse. The conditions on r_c are that the royalty fee be <u>small</u> enough so as not to seem an unreasonable burden to the copier (a psychological judgment), and <u>large</u> enough so that, together with the independent variable N, the total Nr_c pays each publisher a non-trivial additional revenue, <u>and</u> meets the operating costs of the clearinghouse. The conditition on a are that it be <u>small</u> enough not to deviate a large share of receipts to overhead (another psychological judgment). In this model we assume the value of a is 1/10 or less.

Two variables are not within control of the clearinghouse: N, the total volume of national copying of copyrighted pages per year, and $\bar{c}$, the average cost, for all publishers, to publish one page of their publications. $\bar{c}$ is important for fair distribution of fees. N, however, dominates the feasibility and success of the entire clearinghouse operation. Let us see how this is so by estimating the actual dollar flow into the clearinghouse.

In 1966, one of the first estimates was made of total national photocopying of copyrighted works [3]. On the order of twenty billion pages per year were photocopied. Of these, on the order of one billion were copyrighted. A recent estimate of total national photocopying is 78 billion pages per year [11]. Assuming the same ratio (the great bulk of photocopying is of uncopyrighted material, for office and record keeping. Photocopying of copyrighted works is mainly by scientists and students) namely, 5%, there were about 4 billion copies of copyrighted material made in 1975. This represents a growth in annual copyrighted photocopying on the order of 400% in nine years. It is estimated that by 1980, copyrighted photocopying will at least double that in 1975, and by 1986 at least triple it, a growth of 300% in eleven years. Assuming a copyright clearinghouse could be brought into steady-state operation by 1980, a conservative estimate of copyrighted annual copying would be 8 billion pages per year. By 1986, it would be on the order of 12 billion pages per year. Therefore the design parameter N which we assume for the United States is:

	1966	1975	1976	1980	1986
N (in units of 10^9 copyrighted pages per year)	1	4	5	8	12

Figure 3 is a matrix plot of royalty fees r against N (and hence against time). Values and time range of N are as above, the range of r is from 0.2¢ per page copied to 5.0¢ per page copied, and the value of C = Nr ranges from 2 to 600 million dollars. Reasons for the range are discussed below. A rough economic design of the projected copyright clearinghouse can be followed on this matrix. The dashed lines which partition it represent constraints -- values which are not ruled out are in the remaining smaller region between them. A yet smaller matrix region lies within this one, and represents the projected region in which the steady state will be feasible with all constraints met. Hence it is the "self-maintaining operating region" of the proposed clearinghouse.

Royalty fee (r), dollars per copyright page copied	Estimated total pages(N) copyrighted and copied per year in the United States. (multiples of 10^9/year)				
	1966	1975	1976	1980	1986
	1	4	5	8	12
0.002	2	8	10	16	24
0.004	4	16	20	32	48
0.006	6	24	30	48	72
0.008	8	32	40	64	96
0.010	10	40	50	80	120
0.012	12	48	60	96	144
0.014	14	56	70	112	168
0.016	16	64	80	128	192
0.018	18	72	90	144	216
0.020	20	80	100	160	240
0.025	25	100	125	200	300
0.030	30	120	150	240	360
0.035	35	140	175	280	420
0.040	40	160	200	320	480
0.045	45	180	225	360	540
0.050	50	200	250	400	600

Overall cash return (C = Nr), in multiples of $\$10^6$, millions of dollars per year.

FIG. 3 Economic Design of Copyright Clearinghouse, in the Steady State

Region constraints: $r_c \geq r_p$, $Nr_c a \geq$ Clearinghouse budget (B)

$Nr_p(1-a) \geq$ publisher threshold, $0 \leq a \ll 1$, $r_c \leq$ user threshold

To determine the value of the first parameter, r (here called r_p) which would yield a return $C = Nr_p$ sufficiently large, in view of the overall volume N, so that C offers even the largest publisher a strong incentive to join, is, of course, to assign a subjective value. An arbitrary value must be selected. Assume, as shown in the 1966 report, that there are fewer than 1000 publishers (more on the order of 700 in 1966, which we round out to as many as 1000 for reasons discussed below), and that, as shown in the same report and elsewhere, approximately 6% of these publish about 40% of the pages copied. Assume further, that even the largest publisher will be interested in an annual return from copying which he might otherwise not collect, it it amounts to at least \$200,000. This is an arbitrary design figure, obviously open to debate but assumed for illustrating the economic design.

Solving for C (overall cash return) by the simple relation:

$$C \times \frac{0.40}{60} - \$200,000$$

$$C = \$30 \times 10^6$$

A line bordering the C-region corresponding to an annual return of at least \$30 million could be drawn on the matrix, and the r ordinates just below this line would represent the values of r_p to yield this minimum annual return. Since, however, C must equal $N(1 - a)r_p$ rather than Nr_p, in which a is at most 0.1, the value of C which must be exceeded is 10/9 of this amount, or \$33 million per year. The broken line passes above all parts of the matrix equal to or greater than \$33 million. The line shows that in 1966, when overall copying of copyrighted pages was only one billion pages per year, there was no royalty fee under 3.5¢ per page which could (under the criterion used) make it worthwhile for the larger publishers to join. In other words, to satisfy the publisher-return criterion, given that annual volume of copying, it would have been necessary to nearly double the minimum cost of copying a page (without royalty) in order to pay the royalty fee. This would have been unacceptable to most copiers, who would have had to pay about 8¢ per copied page. On the other hand, with the volume projected for 1976 the copier could pay a royalty fee of 0.8¢ per page, for a total of about 5¢ per page, other conditions remaining the same. By 1980 the royalty rate could be reduced to 0.6¢ per page, and by 1986 to as low as 0.4¢ per page, other conditions the same, and yet satisfy the publisher criterion. Thus the effect of larger volume is to maximize the copyright owner's return and also enable reducing the royalty cost to the copier, so that the total cost to him approaches barely more than the copying cost alone [12].

A value of r_c, the actual rate at which the copier's tolerance of a royalty fee begins to change his behavior in making decisions on copying, is also quite variable. A well-to-do copier might not balk at paying 25¢ per page (say 5¢ for the ink, paper and profit on use of the machine, and 20¢ for royalty fee). However, many might consider this proportion of copyright royalty to copying cost unfair, especially in view of the uniform fee for copying any page, whether expensive or cheap to publish. Also, copying takes place over a broad spectrum of activity -- from bulk copying for educational purposes to special copying for advanced research -- but the "center of gravity" is closer to the elementary school than to the corporate research division. It follows that from the copier's point of view, the single, uniform fee should not add very much to what the copier now pays for copying only, without royalty fee.

Consider the present (April, 1976) range of copying fees, by xerox or other means. The rate has gradually come down so that commercial copies are made for as little as 3¢ to 4¢ per page. This includes profit but no royalty fee. A very large bulk of present copying is at the rate of 5¢ per page, with a sharp fall off in volume for higher rates. For the purposes of this rough estimate, let us assume the copier's tolerance is not more than 6¢ per page, overall. That is, if he needs a number of copies he will, with little hesitation, pay 6¢ as readily as 5¢ per page. Beyond this he may cut down on the number of pages, look for a cheaper copying establishment, cheat a little, or show other forms of fee-sensitive behavior. If the actual cost of copying were 3¢ (no profit), a 6¢ rate would allow a total profit of 3¢ per page, to be divided between the supplier of the copying device and the owner of the material copied. Let us assume they divide equally. The copyright owner could under the psychological tolerance threshold principle, receive 1.5¢ per page, but not more. This figure would set the maximum rate which the clearinghouse would charge as the uniform fee on all pages copied under its contracts. The dashed line corresponding to $r_c = 1.5$¢ is shown in Figure 3. Together, the two parameters r_p and r_c enclose an area on the chart within which the clearinghouse can operate successfully, and meet all three constraints (copier's tolerance, copyright owner's incentive, overhead to clearinghouse). It is seen that for 1966 no value of r_p was less than r_c. From 1975 on, there is a widening range for which $r_c - r_p \geq 0$. However, there is one more computation which must be made before we can be reasonably sure that the design will work.

It is not sufficient to require that the clearinghouse take not more than 10% of the total receipts for its self-maintenance. We must also be assured that the actual amount so taken corresponds to or exceeds a total budgeted operating cost of an appropriately staffed and capitalized organization. The staff must perform all its functions -- periodic sampling, collecting fees from larger copying organizations, dispensing stamps or other licenses for individual copiers, disbursing periodically to the approximately 1000 publishers. A large staff is not needed. Assume that in 1980 the job could be done by 250 employees. If the average outlay per employee in a similar organization were on the order of \$50,000 per year this would require a budget of \$12.5 million per year. Solving for the magnitude of C on the basis that

$$0.1\ C = \$12.5 \times 10^6$$

$$C = \$125 \times 10^6$$

Comparing this result with Figure 3, we find that an annual return of \$125 million would first become possible in 1980, but with no latitude in the royalty fee. The fee would have to be the maximum tolerated, 1.5¢ per page copied. By 1986, the same size clearinghouse would have an option of setting the fee between 1.2¢ and 1.5¢ per page copied, and could exercise its long-term policy of reducing the fee. This assumes that in 1986, with projected copying volume increased from 8 to 12 billion pages per year, the same sized staff could handle the job; and the value of the dollar and other parameters remained the same. In practice the revision of copyright fee would take many factors into account. The above exercise does indicate, however, that within the next few years and almost certainly in a decade, the steady-state economics are such that a copyright clearinghouse could become a viable national organ of communication. Since the start-up phase may consume several years, we may conclude that the steady-state self-supporting copyright clearinghouse could become feasible at about the time it would take if we started to organize it now.

B. Fair Distribution: While there are many objectives for sampling, the main one, as stated, is to insure fair distribution of royalty fees to publishers without excessive record keeping and overhead. To obtain a fair sample is not necessarily simple. While the actual population of publishers is not large, a large fraction of these are so rarely copied that they would not "show" or "be resolved" except by huge samples, or very frequent samples.

In either case, the cost would be prohibitive, and defeat the objective. Fortunately, it is unnecessary to take so much data that works of rarely copied publishers will always be represented. An alternative is to truncate the distribution of publishers *versus* copying activity (arranged in descending order) at about the level of activity at which presence or absence from any sample falls below some arbitrary threshold, and treat the below-threshold publishers uniformly and adequately. E.g., each might receive an annual return fixed to be more than they could expect, given no clearinghouse, and therefore a definite incentive to join the clearinghouse. The amount should not be negligible, yet not so large as to tempt non-*bona fides* to join the clearinghouse merely for the sake of sharing the income. An upper limit need not be set on the number of below sampling-threshold publishers, but admission to the clearinghouse should require proper identification.

Allocation of payments among publishers who are genuine, but rarely show on copying samples, would be simple. Let their number be q. Let X be some predetermined fraction of the annual copyright royalty fees, $N(1 - a)r_c$. Then the share of each would be:

$$S(q) = \frac{XN(1 - a)r_c}{q}$$

while the remaining n - q members of the clearinghouse would share

$$(1 - X) \cdot N(1 - a)r_c$$

between them, but not equally. If any publishing house began to show consistently in successive samples, it would automatically transfer from the q to the n - q group. For example, let n = 1000, q = 300, X = 0.04, $N(1 - a)r_c = \$30 \times 10^6$. Then S(q) = \$4000 would go to each of 300 members, while 700 members would receive, on the *average*, \$41,000. Actual annual payments to these latter might range from several million dollars to \$4000.

The main reason for sampling, then, is to deal fairly with the n - q publishers. Complete fairness, however, would not be served by apportioning receipts by numbers of pages copied from the works of each. The number of pages copied should be weighted by average cost to publish a page of the particular publisher's works. We define

P_{ij} = number of pages copied in the ith sample, from works of the jth copyright owner (jth publisher). Evidently $P_{ij} \geq 0$.

N = total number of publishers in the clearinghouse, whether or not all appear in the ith sample.

$$P_i = p_{i1} + p_{i2} + \ldots\ldots + p_{in} = \sum_{j=1}^{n} P_{ij} \qquad (3)$$

P_i is the sum of all pages copied from copyrighted works of all the publishers in the clearinghouse, in the ith sample. If $p_{ij} = 0$, the jth publisher does not appear in the sample.

c_{ij} = average cost to the jth publisher of publishing one page of his total annual page production.

This figure is easily reached by dividing the total cost (not including profit) of publishing all works by pages per year (e.g., a summation over all editions of the total number of pages in a volume multiplied by the number of volumes in the edition is divided into the cost to produce these pages). A publisher, on joining the clearinghouse, would agree to supply these overall figures; if necessary, to show the data on which they are based. Then

$p_{ij}c_{ij}$ = cost of publishing the copied pages in sample i of the jth publisher's works.

$$K_i = p_{i1}c_{i1} + p_{i2}c_{i2} + \ldots = \sum_{j=1}^{n} p_{ij}c_{ij} \qquad (4)$$

K_i represents the cost of publishing all pages copied from all publishers in sample i. Absence of the kth publisher's works from the sample would affect the summation over n to the extent that $p_{ik} = 0$. From the two totals, p_i and K_i, a new average can be obtained:

$$\bar{c}_i = K_i/P_i \tag{5}$$

$\bar{c}_i$ is the average cost per page, of publishing all pages in the ith sample. Its value does not necessarily coincide with any $\bar{c}_{ij}$ of an individual publisher. The relative share of the jth publisher in the return from the ith sample is:

$$\frac{p_{ij}c_{ij}}{\sum_{j=1}^{n} p_{ij}c_{ij}} = \frac{p_{ij}c_{ij}}{K_i} \tag{6}$$

If there are a total of m similar samples per year, the mean relative share of the jth publisher would be:

$$\frac{1}{m}\sum_{i=1}^{m} \frac{p_{ij}c_{ij}}{K_i} \tag{7}$$

Equation (2) expresses the total dollar amount of the annual receipts from royalty fees collected from all sources. Therefore the average amount to be remitted to the jth publisher is:

$$N(1 - a)r_c \cdot \frac{1}{m}\sum_{i=1}^{m} \frac{p_{ij}c_{ij}}{K_i} \tag{8}$$

Formula (8) applies to a full distribution of n publishers (no non-showing members such as discussed above). For a truncated distribution with a fraction X of total annual receipts for publishers allocated equally to q of the n publishers, replace n by (n - q) in formulas (2) through (8), and the average share of the jth publisher ($j = 1,2,\ldots,n - q$) becomes:

$$(1 - X)\cdot N(1 - a)r_c \cdot \frac{1}{m}\sum_{i=1}^{m} \frac{p_{ij}c_{ij}}{K_i} \tag{9}$$

while the share of the national annual receipt by one of the q publishers who divide equally is, as before:

$$\frac{X\cdot N(1 - a)r_c}{q} \tag{10}$$

Summing over the n - q terms in (9) and q terms in (10) for T, the total annual distribution of royalty fees to publishers:

$$T = \sum_{j=1}^{n-q} (1 - X)\cdot N(1 - a)r_c \cdot \frac{1}{m} \sum_{i=1}^{m} \frac{p_{ij}c_{ij}}{K_i} + \sum_{j=1}^{q} \frac{X\cdot N(1 - a)r_c}{q} \cdot (1)^j \qquad (11)$$

$$T = N(1 - a)r_c \sum_{j=1}^{n-q} \frac{(1-X)}{m} \sum_{i=1}^{m} \frac{p_{ij}c_{ij}}{K_i} + X \sum_{j=1}^{q} \frac{(1)^j}{q} \qquad (12)$$

The second term in (12) actually does not involve q, but is simply the term $X\cdot N(1 - a)r_c$. It is shown in (12) as a reminder of the process of fee allocation to publishers insufficiently "resolved" in the copying samples.

The number (m) of samples taken annually during the early years might range from 2 to 6. These samples would be used, not to establish the annual distribution but to update the more elaborate preliminary distribution established in the pilot stage. In the steady state, it is suggested to use reasonably small truncated samples, randomly distributed by location and type of institution, at frequent intervals. Preliminary sampling, without truncation at first, would experimentally determine truncation values for economically viable sample sizes. These matters would require expert adjustments. They would depend on experiments similar, but on a more massive scale, to that reported in Ref. 3, pp. 57-71, in which a sample of photocopying extending over one month was obtained from six libraries distributed by size, function and geopgraphic location. In certain subject areas, such as medicine, records may well exist which can be taken advantage of.

Two figures to be determined during transient and steady state phase of the clearinghouse would be the average and median numbers of pages copied per copying request. These should be considered in setting r_c, since the copier of a small number of pages might be willing to pay several cents per page royalty, while copiers of large numbers of pages might not be able to afford this (or vice versa).

VI. REQUIREMENTS FOR IMPLEMENTATION

Requirements foreseen needed for full implementation of the clearinghouse include:

1. Recognition among all parties concerned in the initial organization of the clearinghouse that copying is a new, valid means of dissemination, here to stay; that it replaces much dissemination by circulation, and requires copyrightwise a new balance of interdependency between publishers and users of information.
2. Recognition that voluntary membership is perhaps (under free enterprise) the most effective way to organize for this new role. With publishers and users equally represented in developing policy for the clearinghouse, neutrality and objectivity can be attained. Participation in governing, receiving benefits and sharing responsibilities of the clearinghouse can motivate all parties through vested interests in its successful operation.
3. Establishment of contracts between the clearinghouse and its members as the legally binding force behind the operation. Recognition that, while such contracts affect all parties, the clearinghouse would bear the brunt of the obligation for carrying out contractual terms.
4. Exemption from antitrust litigation. Since the economic success of the operation would depend on voluntary agreement among its members concerning the setting of a uniform fee charged for all photocopying of copyrighted materials, such agreement would require Congressional exemption from antitrust action.
5. Preliminary funding to cover the start-up phase of clearinghouse operations. While there are many potential sources of capital, ideally it should come from all branches of the information community. The more varied the sources of funds, the wider the support and the less criticism possible.

6. Simplicity of operation. This is essential not only to keep operating costs low but to insure cooperation. Assuming willingness to cooperate as inversely proportional to frequency of interface with the clearinghouse, demands on the time of member organizations and individual users should be kept to a minimum.

7. Recognition of the value of a uniform charge page copied. By setting one fee per page for all types of copied material the individual user's effort would be minimal. He would merely verify that the publisher's works were in the clearinghouse list (a list of publishers, not of their works, i.e., about as long as two or three pages of a telephone directory), and multiply the fee into the number of pages he copies. A copying organization would likewise be relieved of all but nominal bookkeeping. It would merely record the total copies made between fixed dates (a record usually kept), and send one check per period to the clearinghouse for the amount of the fee multiplied into this total. The clearinghouse would be relieved of accounting for individual transactions with the exception of those in the sampling. It would take upon itself all weighting computations based on varying publication cost (Sec. V). However, this would be reduced to a minimum by the publishers, whose sole efforts would consist in supplying their publication list and one figure, average cost per page to publish. The publishers would receive one check per period, automatically weighted by this figure.

8. Recognition of the value of the sampling system in design of the clearinghouse and pilot study. A sampling study should be performed before establishing the clearinghouse in order to determine the initial distribution of royalty funds collected, and other variables. A continuing sample should be made periodically, thereafter, to update the distribution. The aim of sampling, in all cases, is to meet or exceed a desired degree

of reliability, yet maintain a low cost for clearinghouse operations. The information must be accurate, yet obtained within budgetary, physical and administrative limitations. The samples should be taken with the consent of the organizations where sampling is done, yet without disturbing their operations more than absolutely necessary. Confidence in joining and in participating in the clearinghouse by publishers on the one hand, and by libraries and other copying institutions, on the other, depends heavily on the reliability of sampling.

9. Recognition that, although the copyright clearinghouse has been advocated for years, it has only recently (following the model) come within range of self-sustaining feasibility. Now, at a time it is urgently needed to preserve the copyright principle in the face of new technology, technology and systems analysis suggest that it can succeed. Development seems appropriate now.

VII. CLEARINGHOUSE AS FACILITATOR OF NEW COMMUNICATION TECHNOLOGY

At this point let us recall briefly that the basic "villain" which motivates these and other studies is the advance of new technology in use of copyrighted material. An ongoing technological change, as we approach the 21st century, is a trend toward non-paper storage of information; and toward rapid, on-demand replication in (usually expendable) paper and other media. The first is exemplified by electronic storage and computer networks [13]. The second, starting earlier, is shown by evolution of the duplicating (or D) library out of the older circulating (or C) library [4][12]. An economic model of D library on-demand dissemination showed that it obeys the same law of "mass production" as the clearinghouse: with sufficient volume (numbers of pages copied) the cost per copied page declines to a lower limit set by the bare cost of the message material and labor.

The same law applies to the D publisher, who differs from the D library in owning his copyrights. As shown in Figure 4, the D library and copyright clearinghouse are naturally symbiotic [4]. The D publisher, however, also benefits. The clearinghouse-plus-D library offers a main outlet in addition to direct sales, what might be called a copying alternative to edition publication. Not shown on the diagram (but implied) are ordinary C libraries (libraries that employ copying equipment are partial D libraries); ordinary (non-D) publishers, bookstores; various classes of individual and institutional copyright owners and users. All of these can make use of the clearinghouse by contract or action under a contract. The clearinghouse is an obvious tool for interlibrary loan, for development of interlibrary and intercomputer networks. Its operations need not be confined to paper media [13], but can be expanded to cover any medium of interchange in which a copyright fee is involved. Thus the diagram represents a broad universe of future dissemination of copyrighted material. On-demand dissemination tends to conserve resources, eliminate duplicate storage in computers, library stacks, publishers' warehouses. If this kind of reorganization is to be interfaced with free enterprise, the copyright clearinghouse could become an essential link in the new technology of communication.

VIII. FURTHER DISCUSSION AND CONCLUSIONS

The proposed copyright clearinghouse, like any other attempted solution to a problem which has no unique solution, must balance its strong and weak points against those of other solutions. This was attempted in breadth by the survey of last year's class [5].

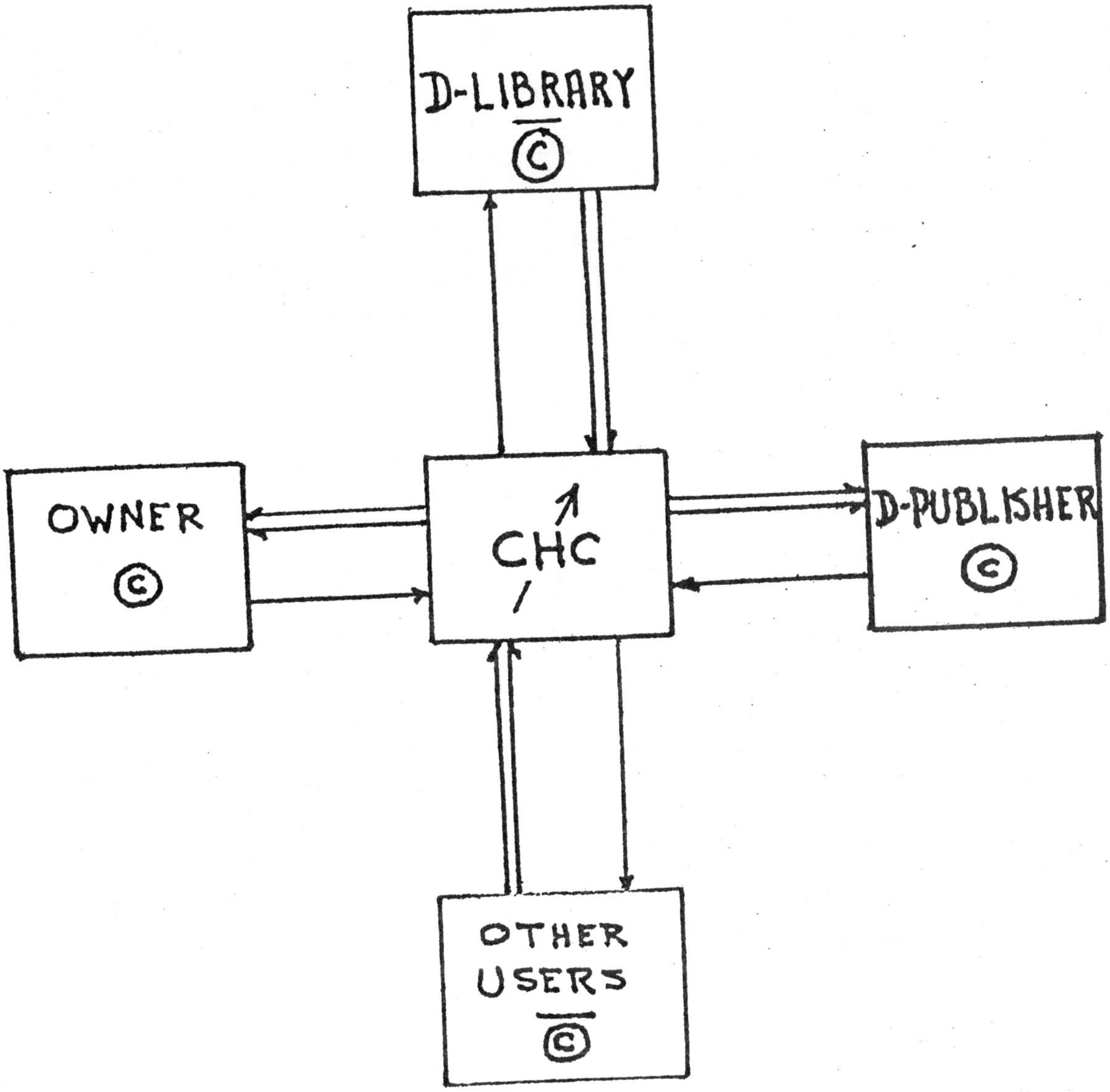

Fig. 4 COPYRIGHT CLEARING HOUSE AS FACILITATOR FOR ON-DEMAND COPYING OF COPYRIGHTED MATERIAL

© COPYRIGHT OWNED

$\overline{©}$ COPYRIGHT USED, NOT OWNED

⇉ COMPENSATION

→ RIGHTS TO COPY

↗ SWITCHING FUNCTIONS

L.B.H.
12-19-75

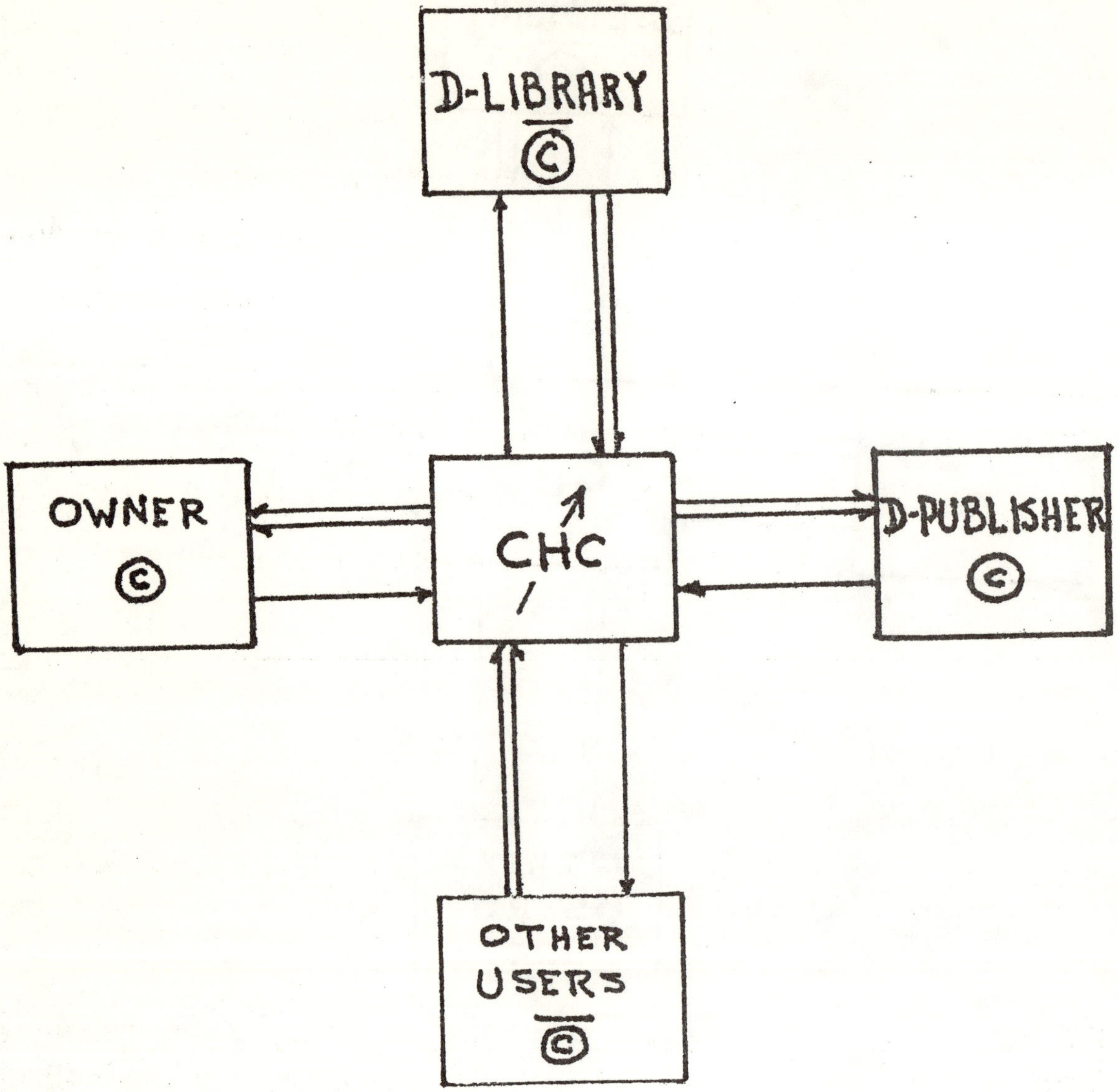

Fig. 4 COPYRIGHT CLEARING HOUSE AS FACILITATOR FOR ON-DEMAND COPYING OF COPYRIGHTED MATERIAL

© COPYRIGHT OWNED

© (overlined) COPYRIGHT USED, NOT OWNED

⇉ COMPENSATION

→ RIGHTS TO COPY

↗ SWITCHING FUNCTIONS

L.B.H.
12-19-75

In this final section some advantages and disadvantages are reviewed.

ADVANTAGES

1. The major advantage of this solution to publishers would be receipt of a fair return for the photocopying of copyrighted material. The solution would guarantee receiving something where there is presently nothing, and would do so with a minimum effort or expense by participating publishers.
2. The chief advantage to users would be unlimited, on-demand copying, without time loss in seeking permissions, with protection against infringement suit, at minimum effort or expense.
3. The major advantage to the public would be introduction of a new system for communication by copyrighted work which, organized to deal with the photocopying problem, is capable of adaptation to any medium in which a copyright fee is involved.
4. The clearinghouse would introduce a number of technical innovations. Among these are a uniform fee per copied page, a frequently updated random sampling of copying of copyrighted works, with minimum disturbance to using or publishing communities; voluntary membership in the clearinghouse, with compelling incentives to join or to use.
5. A feature of the systems design is use of value thresholds in setting the uniform copying fee, minimum publisher's incentive, proportion of proceeds for maintaining the clearinghouse. Another is partition of the population of publishers between those resolved by the sample and those unresolved. The latter are compensated equally, the former through a measure based jointly on volume and publication cost/page.

DISADVANTAGES

1. The authors of copyrighted works have not been specifically included in the contract between publishers and the clearinghouse. It is assumed that a share in returns from copied works will be included in publisher-author contracts. Feasibility of the clearinghouse depends on its dealing, not with copyright owners, of whom there are a large number, but with publishers, who number less than a thousand. In achieving this the clearinghouse concept assumes that the original creator of a copyrightable work would retain the right to bargain with publishers for original publication and royalty schedules, and that D-publication constitutes such original publication. Bargaining, however, may be affected to some degree by clearinghouse operations. Unlimited photocopying of their works may affect returns to authors from subsidiary rights, e.g., book club and serialization rights. To a considerable extent this and similar objections on the part of the author can be overcome by including in the clearinghouse governing board ex officio representatives of authors' organizations, and by changes in author-publisher contracts.
2. Success of the clearinghouse depends on volume, and this requires exemption from antitrust liability. The Congress may not necessarily approve such a measure even for a non-profit corporation designed to serve the nation and to be self-sufficient.

CONCLUSIONS

In the course of working it out, no doubt other disadvantages and advantages would (hopefully, will) be found. On balance, advantages appear far greater. It should be recalled, in reviewing, that the proposal is not new. Some form of copyright clearinghouse has been advocated for nearly two decades. At the first CICP Convention in 1959, there was support for the idea by libraries, educators, suppliers of copying equipment; opposition by some publishers, on the basis that it would cost far too much to record transactions, yield too little in benefits. In retrospect the publishers were right--at that time. The present model has eliminated all detailed accountability except in the taking of samples; on the assumptions for necessary volume of transactions, there has been no time until the present when the clearinghouse could have paid handsome benefits to creators and publishers, become self-sustaining, yet charged reasonable (decreasing with increase in volume) copyright fees to users. Assuming a minimum organization time of three years, by starting now (mid 1976) the clearinghouse could, on the assumptions of this study, reach in 1980 a volume capable of satisfying all requirements; if not then, almost certainly within a decade. It is concluded that since the clearinghouse would offer a versatile service not now available for interchanging permissions and fees, its two stages should be intensively studied by concerned parties and agencies. With the aid of much thought, organization skill and hard work, the clearinghouse could become an instrument for freeing-up the sticky national problem of protecting creativity, free enterprise and use of advanced technology, i.e., for maintaining the copyright principle in an age of advancing communication.

IX. REFERENCES

[1] U. S. Senate, Copyright Law Revision; Senate Calendar No. 460, November 1975, 94th Congress, 1st Session.

[2] U. S. Senate, Senate Copyright Bill S 22, January 1975, 94th Congress, 1st Session.

[3] Sophar, Gerald J. and Heilprin, Laurence B. Determination of Legal Facts and Economic Guideposts with Respect to the Dissemination of Scientific and Educational Information as It Is Affected by Copyright-a Status Report. Committee To Investigate Copyright Problems Affecting Communication In Science And Education, Inc., U. S. Dept. of Health, Education, and Welfare, Office of Education, Bureau of Research. December, 1967. NTIS No. PB 178-463, ERIC No. ED 014-621.

[4] Heilprin, Laurence B. "Technology and the Future of the Copyright Principle". Phi Delta Kappan, Jan. 1967; American Documentation, 19, No. 1, Jan. 1968; reprinted, Key Papers in Information Science, Arthur W. Elias, ed., American Society for Information Science, Washington, D. C., 1971.

[5] The Class in Operations Research for Library and Information Science: LBSC 858H. Proposed Solutions to the Problem of Photocopying Copyrighted Material: A Survey. College of Library and Information Services, University of Maryland, 30 April 1976. To be published.

[6] Study Group of The Committee To Investigate Problems Affecting Communication In Science And Education, "First Annual Report of The Committee To Investigate Problems Affecting Communication In Science And Education." Bulletin of the Copyright Society of the U.S.A., Vol. 10, No. 1, October, 1962.

[7] Heilprin, Laurence B., Working Paper on the Feasibility of a Copyright Clearing House. Committee to Investigate Copyright Problems Affecting Communication in Science and Education, Inc., Washington, D.C., June 26, 1963.

[8] Heilprin, Laurence B., "An Experimental Copyright Moratorium." Final Report to the American Society for Testing and Materials, July 1970. NTIS No. PB 195 941; ERIC No. ED 045-110.

[9] Nemeyer, Carol A., Scholarly Reprint Publishing in the United States. R. R. Bowker, 1972.

[10] Fry, Bernard M. and White, Herbert S. Economics and Interaction of of the Publisher - Library Relationship in the Production and Use of Scholarly and Research Journals. U. S. National Science Foundation, Office of Science Information Service, November, 1975.

[11] "What Hath Xerox Wrought", Time Magazine, March 1, 1976, pp. 69-70.

[12] Heilprin, Laurence B., "The Economics of 'On Demand' Library Copying." National Microfilm Proceedings, Vol. 11, 1962, pp. 311-339.

[13] Bamford, Harold, E. Jr., "Concept for Applying Computer Technology to the Publication of Scientific Journals." Journal Washington Academy of Science, Vol. 62, No. 4, 1972.

SELECTED, ANNOTATED BIBLIOGRAPHY ON PHOTOCOPYING OF COPYRIGHTED WORKS, 1970-1976, WITH A SPECIAL SUB LIST ON THE WILLIAMS AND WILKINS vs. US.S CASE

An Independent Study in Library and Information Science for Course LBSC 859

College of Library and Information Services
University of Maryland
College Park, Maryland

September, 1976

by

Peri L. Schuyler, Graduate Student

Jerry Kidd, Faculty Advisor

This bibliography is compiled in fulfillment of the course work for LBSC 859. Its genesis was conceived in the course of an earlier study, LBSC 858H, Operations Research in Library and Information Science.

It is hoped that bringing together in one place as many citations as possible will benefit researchers on the period immediately prior to and including the passage of the Copyright Revision Act of 1976.

The bibliography is divided into two parts. The first and largest covers citations on photocopying and copyright law from 1970 through 1976, arranged alphabetically by author, and chronologically "within author." Three notations are provided, when applicable: news--brief articles (which are not annotated); international--articles covering international copyright (in some cases an individual country is named, following the word "international"); and Williams and Wilkins--citations relevant to that case.

The second part segregates the material on the Williams and Wilkins vs. U.S. case. These citations appear in both Parts I and II but are annotated only in Part II.

Of the approximately 275 citations nearly half are annotated. A few articles which deserve annotation are cited only, because of unavailability of the source of or time to search.

AAP: a progress report to the industry. Publishers' Weekly, 200(15): 30-34; October 11, 1971. (news)

AAP on photocopying: METRO's free copying plan. Library Journal, 100: 813; May 1, 1975. (news)

An act for the general revision of the copyright law, Title 17 of the United States Code, and for other purposes. Senate Bill 1361, ninety-third Congress, second session. Washington, D.C.: Congress of the U.S., House Committee on the Judiciary, September 12, 1974. 81p.

This is the text of the Senate bill for copyright revision put before the 93rd Congress. In particular, the section on exclusive rights, now includes in the area on limitations, statutory reference to the scope of fair use and reproduction by libraries and archives. Also included in the area of exclusive rights are specifications regarding the interpretation in conjunction with computers and other information systems.

American Association of Law Libraries. 65th Annual Meeting, July 2-6, 1972. General session on copyright. Law Library Journal, 65(4): 443-53; November, 1972. (Williams & Wilkins)

Armada-strength coalition launched against pirates of copyrighted works. American Libraries, 6: 29; May, 1975. (news)

Arntz, Helmut. Reprography and copyright. Copyright, 11(4): 95-102; April, 1975. (international)

The author reviews the recent, rapid growth of technology in the development of various reprographic processes and assesses the impact of this overwhelming advance on the owners of copyrighted material. The practices with regard to uses of copied material for scientific purposes and for industry, the role of micro-publishing, and the effect of computers on copyright protection are discussed at length. He concludes that in keeping with the new technology, entirely new methods of dealing with this problem must be sought and proposes in this light, a solution whereby a surcharge or tax would be attached directly onto the sales purchase price of copying equipment, which would act as a royalty payment when distributed back to the copyright owners. The author cites that this approach has already been carried out successfully in sales of tape recorders in certain countries.

Association of American Publishers. _Copyright permissions: a guide for non-commercial use_. Washington, D.C.: The Association, 1975. 6p.

This brief report outlines procedures for obtaining permission from copyright owners to duplicate copyrighted materials. The reproductions must be for non-commercial use, and the guide is aimed at professionals in fields most likely to serve as intermediaries for this purpose--librarians, teachers, researchers, etc.

Australian Copyright Council. _Photocopying and the law: a guide for librarians and teachers and other suppliers and users of photocopies of copyright work_. Sydney, Australia: The Council, 1970. 7p.
(international, Australia)

This is a concise guide to some of the ranges of permissible photocopying uses of copyrighted material under the Copyright Act of 1968 in Australia. Oriented toward librarians and teachers, the guide discusses areas of private use, research purposes, published criticism or review. The feeling is that the law is not overly restrictive but that there must be some limit set on the practice of wholesale, unlicensed copying.

-B-

Bachrach, Morton W. OE's new copyright policy. _American Education_, 6(7): 28-29; August-September, 1970.
(news)

Baker, John F. & Wagner, Susan. AAP's fourth annual meeting; growing confidence in dealing with a multitude of publisher problems. _Publishers' Weekly_, 205(20): 37-47; May 20, 1974.

Banki, Peter. Reprographic reproduction of works protected by copyright: Washington - June, 1975. _Australian Copyright Council Bulletin_, no. 15: 1-15; September 24, 1975.

The author attended the 1975 Washington meeting on the problems of reprographic reproduction as the representative of the Australian Copyright

Council. The meeting was jointly sponsored by UNESCO and the World Intellectual Property Organization (see Subcommittee of the Intergovernmental Copyright Committee ..., 1975), and the report issued from the meeting recommended that each member state resolve these problems individually.

Barber, Elizabeth. American copyright confusion. <u>The Author</u>, 86(2): 49-51; Summer, 1975. (news)

Barker, Ronald E. <u>Photocopying practices in the United Kingdom</u>. London: Faber & Faber, 1970. 104p. (international, United Kingdom)

A four-part study is undertaken based on a questionnaire (reproduced in the book) sent to two thousand members of ASLIB covering: 1) exposition of current copyright law in the United Kingdom and its interpretation with respect to photocopying; 2) an analysis of current photographic reproduction practices; 3)considerations to be thought of in evaluating these practices; and 4) the author's conclusions. It is felt that adjustment of the economics of authorship and publishing should be made without delay to meet the demands of a new era of publishing.

Barnhill, P. To copy or not to copy: that is the question. <u>Kentucky Library Association Bulletin</u>, 38: 13-14; Fall, 1974. (news)

Beard, Joseph J. The copyright issue. In: <u>Annual Review of Information Science & Technology</u>, vol. 9, ed. by Carlos J. Cuadra, et al. Washington, D.C.: American Society for Information Science, 1974. 381-411.

The author begins with a review of the origins of copyright law from ancient times through English Law and the beginnings of the United States copyright law. He then turns to the doctrine of fair use and examines many of the cases brought to the courts, illustrating the difficulties in interpretation of this doctrine. Of the copyright issues of today, he focuses on two: 1) reprography, and 2) computer-based information systems. The reprographic issues are discussed in sections on the author/publisher and the educator and the author/publisher and the librarian. He concludes that the "heart of the problem, whether one is talking about reprography or computers, is marketing." Methods must be established for selective availability of information in the same way that purchasers need not buy a whole steer in order to eat steak.

Bender, Ivan. Interview: "Unauthorized copying": a bubbling issue. *Nation's Schools*, 91(2): 69-70; February, 1973. (news)

Bender, Ivan. Copyright: chaos or compromise. *LJ/SLJ Previews*, 3-5; November, 1973. (news)

Bender, Ivan & Engler, David. When is it legal to duplicate? *Media and Methods*, 11(5): 42-46; January, 1975.

The authors present a review of the Copyright Law and a discussion of the judicial doctrine of "fair use", not yet a statutory law (at time of article publication). The article reviews the four conditions which must be met for "fair use", the standards to be met in making reproductions, and ends with several examples of fair and unfair use with explanations.

Benjamin, Curtis G. Who will save the scholarly book? *American Scholar*, 41(2): 212-21; Spring, 1972.

Benjamin, Curtis G. Reprint editions: some pertinent questions. *Publishers' Weekly*, 202(19): 22-23; November 6, 1972. (news)

Benjamin, Curtis G. A hard look at the new Williams & Wilkins decision; the Court of Claims ruling presents a somber prospect for a wide variety of specialized publishers. *Publishers' Weekly*, 205(10): 33-34; March 11, 1974. (Williams & Wilkins)(news)

Benjamin, Curtis G. Photocopying: world bodies decide against international guidelines. *Publishers' Weekly*, 208(7): 28; August 18, 1975. (international) (news)

Benjamin, Curtis G. Regulation of photocopying: a world-wide quandary. *Library Journal*, 100(15): 1481-83; September 1, 1975. (international) (news)

Benson, Stuart W., III. Technological piracy in the reprographic revolution. *University of Pittsburgh Law Review*, 36: 153-73; Fall, 1974. (Williams & Wilkins)

Berkowitz, Linda L. Copyright law - fair use - photocopies: Williams & Wilkins Co. v. the United States. New York Law Forum, 20(4): 819-31; Spring, 1975. (Williams & Wilkins)

Berry, J.N. Copyright: from debate to solution? Library Journal, 100(15): 1459; September 1, 1975. (news)

A bill for the general revision of the Copyright Law, Title 17 of the United States Code, and for other purposes. Senate Bill 22, Ninety-fourth Congress, first session, calendar number 460. Washington, D.C.: Congress of the United States, Senate Committee on the Judiciary, January 15, 1975 (as amended November 20, 1975). 158p.

The text of the amended Senate bill for copyright as put before the 94th Congress is presented. Sections 107 and 108 on fair use and reproduction by libraries and archives for the first time attempt to statutorily define the scope and interpretation of fair use and the extent to which photocopying of copyrighted material is allowable under this limitation to the exclusive rights of authors.

Blackburn, Robert H. Photocopying in a university library. Scholarly Publishing, 2: 49-58; October, 1970. (international, Canada)

This is a report of a sampling study carried out for a two-week period from 23 February to 8 March 1970 at the University of Toronto Library to determine the extent and type of photocopying practices. The conclusions are: 1) that the library photocopying deals mainly with serial publications where authors derive no royalties and publishers do not meet public demand for more than a short while after each issue appears; 2) that the amount of copying done in the library is only a drop in the bucket compared with the rest of the university or the city; 3) libraries need to make photocopies for purposes of research and private study, and efforts to establish regional and national specialization of collections and coordination of information systems would be negated if libraries stopped photocopying out of fear of breaking the law; 4) that unauthorized production of copyrighted editions of a work for sale or classroom use or general distribution without permission is illegal; and 5) in the author's opinion it is a mistake to equate photocopying and computer storage of data in relation to their impacts on publishing.

Blackwell, Thomas E. The law of copyright and the fair use doctrine. Journal of College and University Law, 1(3): 222-29; Spring, 1974.

Library photocopying problems regarding copyrighted materials are analyzed both in a historical perspective and in light of the Williams & Wilkins case, especially in connection with the difficulties in interpretation of the fiar use doctrine. The author discusses in detail the proposed revisions of the copyright law now before Congress, the prospects for a compulsory licensing system, and concludes with a recommendation that the federal government return compensation to publishers for the educational use of copyrighted materials.

Brewer, Karen. A healthy approach to copyright problems. Publishers' Weekly, 202(11): 21-22; September 11, 1972. (news)

(Refers to Doebler, Publishers' Weekly, 202(4): 52-54; July 24, 1972)

Breyer, Stephen. The uneasy case for copyright: a study of copyright in books, photocopies, and computer programs. Harvard Law Review, 84(2): 281-351; December, 1970.

An extremely lengthy analysis of the cases is presented in favor of copyright protection on a "moral basis" with countermanding arguments on an economic basis. Although the slant of the work is primarily aimed toward a discussion of copyright protection in books, part three deals at some length with the extension into areas of photocopying in periodicals, and protection of computer programs. It is Professor Breyer's opinion and conclusion that contrary to most publishers' claims, there are sound economic arguments in favor of removing copyright protection, that the "general case for copyright protection is weak", that the protection period should not extend beyond fifty-six years, that individuals should be allowed to make single copies of articles, and portions of books without permission, that individuals should be allowed to store copyrighted materials in computers for research purposes without permission from owners, and that computer programs should not be copyrighted at present.

Breyer, Stephen. Copyright: a rejoinder. UCLA Law Review, 20(1): 75-83; October, 1972.

This is the third in a series of communications begun with Professor Breyer's original article (see above Breyer, 1970). That paper engendered a response by Barry Tyerman (see Tyerman, 1971) to which Professor Breyer offers rebuttal in the present article.

Brooks, B.C. Photocopies vs. periodicals: cost-effectiveness in the special library. Journal of Documentation, 26: 22-29; March, 1970.

Bungeroth, Marc R. Copyright - photocopying as fair use. Boston College Industrial and Commercial Law Review, 16(1): 141-50; November, 1974. (Williams & Wilkins)

Bush, George M. Technology and copyright: annotated bibliography and source materials. Lomond Systems, Inc.: Mt. Airy, Md., 1972. 454p.

This book is divided into two parts. Part one contains about three hundred annotated bibliographies classified under various subject areas and arranged by date from pre-1966 until 1972 (eleven citations in all from 1972). Part two contains copies of nineteen relevant articles from the literature dealing with technology as it affects the copyright law and its interpretation. Of particular interest are the citations in part one in sections D and F through I dealing with reprography, microforms, fair use, library and educational topics.

-C-

Cambridge Research Institute. Omnibus copyright revision: comparative analysis of the issues. Washington, D.C.: American Society for Information Science, 1973. 280p.

This is an extremely thorough and useful book examining in depth the whole scope of the copyright revision bill. In three sections, the book deals first with the historical background of copyright in the United States, internationally, and with earlier attempts at revision of the law. Part two discusses each of the individual issues in the revision section by section in the general form of a statement concerning the existing doctrine, an analysis of the areas leading to the decision to revise, or a discussion on the impacts of possible revision and a comparison of the two, and a conclusion. Part three is in tabular form consisting of a section by section comparison of the present law with the revised version. Column one lists the paragraph in the old law, column two the corresponding section in the new, column three is a comparative analysis summarizing the differences between the two. Of particular interest in this instance are the paragraphs in parts two and three analyzing the previously uncodified statements on fair use (Section 107 revised) and photocopying in libraries (Section 108 revised).

Carnahan, William H. Fair use and the libraries. *Air Force Law Review*, 17(3): 41-51; Fall, 1975. (Williams & Wilkins)

Carnahan, William H. The copyright dilemma in the classroom. But can it be done? *Air University Review*, 23(2): 42-49; January-February, 1972.

Carnahan, William H. Copyright in our realm of learning. *The College Counsel*, 71(1): 421-29; 1972.

Catterns, David. "The Americans, Baby" by Moorhouse: an Australian story of copyright and the new technology -- The Thirteenth Annual Jean Geiringer Memorial Lecture on International Copyright Law. *Bulletin of the Copyright Society*, 23(4): 213-32; April, 1976. (international, Australia)

The author takes his title from the title of a collection of short stories by Frank Moorhouse, of which the making of two copies of one of these (approximately ten pages long) was the subject of a 1975 infringement case in Australia, The University of New South Wales v. Moorhouse and Angus & Robertson (Publishers) Pty Ltd. The Supreme Court of New South Wales, while granting the infringement had occurred, held that the University was not liable; this ruling was overturned by the High Court of Australia. The case was instigated by the Australian Copyright Council to test the issue of authorization. The author discusses the Australian approach to reprography with respect to the Moorhouse case, the "uneasy case for fair use" both in Australia and in the United States in relation to Williams & Wilkins, a blanket scheme of compulsory licensing, and methods for assessment of the amount of photocopying done, and means for distributing royalties accruing from such copying.

Catterns, David. University of New South Wales v. Moorhouse and Angus & Robertson (Publishers) Pty Ltd. Australian Copyright Council Bulletin, no. 16: 1-8; September 24, 1975. (international, Australia)

This is a case note on the suit initiated by the Australian Copyright Council, University of New South Wales v. Moorhouse and Angus & Robertson (Publishers) Pty Ltd., in order to test the responsibility for liability in unsupervised photocopying of copyrighted material. The author reviews Australian copyright law and proposes a blanket scheme for compulsory licensing with suggestions for administration and royalty collection.

Caughran, Alex. The teacher and the copyright law. Journal of Reading, 17(1): 8-15; October, 1973.

Cohen Jehoram, Herman. Zehn Cents Urhebervergütung je Photokopie in den Niederlanden. Gewerblicher Rechtsschutz und Urheberrecht, Int. Teil, no. 5: 161-64; May, 1975. (international, Netherlands) (in German)

This is a German language article analyzing a 1974 amendment to the Netherlands copyright law expressly directed toward the regulation of photocopying and other reprographic techniques. Generally, these regulations establish a charge of approximately ten cents per page to administrators, libraries, institutions of higher learning for reproduction of copyrighted materials, and to other industrial enterprises for reproduction of scientific and technical material. Reproductions from text books for use in primary and secondary schools are charged at a lower rate of approximately two and a half cents per page.

Conference on resolution of copyright issues considers library photocopying problem. FLC Newsletter, 82: 8-9; March, 1975. (news)

Copying in schools; P.A. offers draft license as basis for discussion. The Bookseller, no. 3644: 2180-87; October 25, 1975.
(international, Great Britain)

The Publishers Association of Great Britain has submitted a draft for a blanket licensing system as a basis for discussion in an attempt to find a solution to the problem of obtaining compensation to authors and publishers for photocopies of text books in use in schools. They propose a fee for each participating school based on the number of students per school.

Copyright, computers, and compulsory licensing. Rutgers Journal of Computers and the Law, 5: 149-69; 1975.

Copyright conference considers photocopying problem. Library of Congress Information Bulletin, 34: 195-96; May 16, 1975. (news)

Copyright--fair use--photocopying and distribution of copyrighted journal articles by government library constitutes actionable infringement. Vanderbilt Law Review, 25(5): 1093-107; October, 1972.
(Williams & Wilkins)

Copyright Law Revision. Hearings before the Subcommittee on Patents, Trademarks, and Copyrights of the Committee on the Judiciary, United States Senate. Ninety-third Congress, First session. Washington, D.C.: Congress of the U.S., House Committee on the Judiciary, August, 1973. 682p.

This is a transcript of the hearings held to consider legislation (S. 1361) for the purpose of copyright revision. The text is included as well as testimony from witnesses on five selected copyright issues: 1) library photocopying; 2) general educational exemptions; 3) CATV royalty schedule; 4) religious broadcasting exemption; and 5) CATV coverage of sporting events. The appendix has forty-nine additional statements from members of publishing, library, broadcasting, and educational associations as well as members of Congress.

Copyright law revision. Report together with additional and minority views (to accompany S. 1361). Ninety-third Congress, second session, calendar number 946. Washington, D.C.: Congress of the U.S., Senate Committee on the Judiciary, July 3, 1974. 229p.

The Senate Committee on the Judiciary presents its amendments to the bill (S. 1361) for the general revision of the Copyright law along with its analysis and recommendations. Briefly, the bill provides for general revision, establishment in the Library of Congress of a National Commission on New Technological Uses of Copyrighted Works, and adds protection for ornamental designs on useful articles.

Copyright law revision. Report together with additional views (to accompany S. 22). Ninety-fourth Congress, 1st session, calendar number 460. Washington, D.C.: Congress of the U.S., Senate Committee on the Judiciary, November 20, 1975. 169p.

This is the report of the Senate Committee on the Judiciary concerning general revision of the Copyright Law, which offers an amendment in the form of a substitute and recommends that the bill as amended should be passed. It includes in the first forty-six pages all sections of the present bill to be substituted and then presents a section by section analysis of the revision. Pages sixty-one to seventy-one discuss sections 107 on fair use and 108 on reproduction by libraries and archives.

Copyright Office and the National Commission on Libraries and Information Science, cosponsors. Conference on resolution of copyright issues. Washington, D.C.: November 16, 1974. 109p.

Copyright Office and the National Commission on Libraries and Information Science, cosponsors. Conference on resolution of copyright issues, 3d. Washington, D.C.: April 24, 1975. 103p.

Copyright: P.A. submission to the Whitford Committee. The Bookseller, no. 3584: 1620-22; August 31, 1974. (news) (international, Great Britain)

Copyrights--infringement by photocopying. Texas Law Review, 51(1): 137-48; December, 1972. (Williams & Wilkins)

Cornish, W.R. Photocopying, libraries and the Copyright Law of Great Britain. International Journal of Law Libraries, 4(1): 26-31; March, 1976. (international, Great Britain)

The author discusses the 1956 Copyright Act of Great Britain with respect to the three provisions that most directly affect photocopying in libraries: 1) infringement by substantial copying; 2) fairdealing (comparable to the United States "fair use") for private use and research; and 3) special exemptions for certain libraries listing the three conditions under which photocopying is permitted: conditions under which single-copy photocopies of periodical articles are allowed, of portions of other published works, and of supplying copies of works to other libraries. He suggests that of all the solutions currently extant, one more likely to be acceptable is that of a blanket licensing system granted to regular users of their material by copyright holders.

Court ruling threatens photocopying. Chemical & Engineering News, p. 14; March 6, 1972.
(news)

Cox, Henry B. The law and the manuscripts curator. Drexel Library Quarterly, 11(1): 72-86; January, 1975.

This article is a study of the law as it affects archivists and curators of manuscript materials. Several sections are outlined with regard to literary property rights, gifts, thefts of documents, etc., but of interest here is the part dealing with the photocopying of manuscripts. In raising the question of whether reproducing manuscripts is covered under common law, the author states that "comment...may well revolve around what a court is likely (author's emphasis) to do as opposed to what it is empowered by... existing legal principles to do in fact." He feels that in light of the Court of Claims decision in the Williams & Wilkins case ruling for the defendants, NLM and NIH libraries, photocopying of manuscripts would likely be considered fair use by other courts.

Crabb, Geoffrey. Copyright and schools: a special case? Times Educational Supplement, 3123: 55-56; April 4, 1975.
(news)
(international, Great Britain)

Crabb, Geoffrey. Copyright and the resource centre. Programmed Learning and Educational Technology, 12(3): 191-98; May, 1975.
(international, Great Britain)

The article is divided into three parts: 1) an analysis of the 1956 Copyright Act in Great Britain with emphasis on the areas relevant to its impact on educational institutions; 2) discussions on copying dealing with substantial use, fair dealing (comparable to United States "fair use"), educational use, libraries, and other exceptions, including a summarizing paragraph related to educational use of copyright material. This latter paragraph outlines the changes in educational methods brought about by recent technological advances, and the difficulties coinciding with these and the current law; and 3) the work of the Council for Educational Technology in attempting to find solutions to these problems.

Culliton, Barbara J. Photocopying habit in jeopardy. Science, 176: 265; April 21, 1972.
(Williams & Wilkins)(news)

-D-

Dahlmanns, Gerhard J. Reprography and copyright: the Joint Meeting of the Committees of the Universal and Berne Copyright Conventions. International Journal of Law Libraries, 2(2): 55-63; July, 1974.
(international)

The author reports on the events, agenda and resolutions of the twelfth session of the Intergovernmental Copyright Committee. He describes the previous history, the deliberations in Paris, other items on the agenda and reproduces the draft resolutions of the committee which finally amounts to putting off a decision until the thirteenth session, recommending that the General Conference of UNESCO do likewise, and appointing a subcommittee to examine the matter further.

Dembart, Lee. Photocopying splits publishers, libraries. New York Times, at p. 30, columns 1-4 (city edition); October 29, 1974.
(Williams & Wilkins)(news)

Derenberg, Walter J. U.S. Senate passes omnibus revision bill (S. 1361) and interim legislation. Bulletin of the Copyright Society, 22(1): 1-9; October, 1974.

The author presents a general discussion on the Senate passage of S. 1361 with commentary on the floor debate among the senators, and the various proposed amendments and their potential outcomes.

De Sanctis, Lorenzo. In tema di reprografia e diritto d'autore. Il Diritto Autore, 45(1): 35-45; January-March, 1974. (in Italian)
(international, Italy)

The paper presents a review of efforts on an international scale to solve the problems associated with the photocopying of copyrighted works and the rights of authors. This is followed by reference to the sections of Italian copyright law pertinent to reprography.

Desjeux, Xavier. Photocopying and copyright. Copyright, 9(3): 51-56; March, 1973. (international, France, Europe)

This study is broken into four sections: 1) an analysis of the French copyright law which states that the right of reproduction belongs to the author, with an exception for private use, emphasizing that broad or lax interpretation of this private use exception is a growing threat to authors; 2) teachings of comparative law with discussion of photocopying and copyright law in Scandinavia, Anglo-Saxon countries and the Federal Republic of Germany; 3) the guidelines provided by international law in an effort to strike a balance between protection of the rights of copyright owners and safeguarding the continued flow of information, and urging that national laws are the appropriate source of such legislation; and 4) the three situations which must be accounted for in determining a solution to the problem: a) photocopying by firms for-profit which would require author consent and where agreement should be sought for between publishers and industrial representatives; b) photocopying by non-profit institutions for educational and research purposes where compulsory licensing may be employed, and collecting agencies, public or private, should be established; and c) photocopying by individuals for personal use, where a fee attached at time of purchase would be affixed to equipment used for this purpose, such fee to be paid to the collecting agency mentioned in (b).

Diamond, Sidney A. Split Supreme Court decision leaves photocopying issue clouded. <u>Advertising Age</u>, 46(19): 50; May 12, 1975.
(Williams & Wilkins)(news)

Dittrich, Robert. Die Vervielfältigung zum eigenen Gebrauch. <u>Archiv für Urheber-, Film-, Funk-, und Theaterrecht</u>, 64: 33-52; 1972. (in German)
(international, Austria)

Dittrich, Robert. Zur urheberrechtlichen Beurteilung von Minzkopierautomaten nach osterreichischem Recht. <u>Gewerblicher Rechtsschutz und Urheberrecht</u>, Int. Teil, nos. 6/7: 257-60; June/July, 1973. (in German)
(international, Austria)

The author evaluates some of the questions arising from the use of coin-operated copying machines with regard to the statutes of Austrian copyright law. The problems of responsibility for possible infringement, third party copying, private use, etc., are analyzed.

Dock, Marie-Claude. The UNESCO International Copyright Information Centre. <u>Information: News and Sources</u>, 7(3): 93-95; March, 1975.
(international)(news)

Doebler, Paul. IIA discusses the copyright dilemma. <u>Publishers' Weekly</u>, 202(4): 52-54; July 24, 1972. (Williams & Wilkins)(news)

Doebler, Paul. Proprietary rights problems confront micropublishers. <u>Publishers' Weekly</u>, 203(12): 52-54; March 19, 1973. (news)

Doebler, Paul. New computerized services provide speedy (but not free) copies on demand: McGraw-Hill and Institute for Scientific Information offer rapid retrieval of huge quantities of published information without infringing copyright. <u>Publishers' Weekly</u>, 206(21): 34-36; November 18, 1974.

Although a brief article, this offers two publishers' approaches to the "on-demand" provision of journal articles to users. The Institute for Scientific Information, through its OATS (Original Article Tear Sheet) service and a hook-up with Lockheed's or System Development Corporation's time-sharing facilities, answers on-line requests for journal articles; they either tear the article from one of their subscription journals (no infringement because no copy) or provide a photocopy (for which a royalty is paid to the publisher). McGraw-Hill Bookstore via Lockheed's time-sharing facility is linked to the National Technical Information Service and can perform searches of the NTIS data base. Many NTIS publications are in stock, but if not, may be ordered on-line directly from NTIS.

Dorffeldt, Siegfried. Urheberrechtliche Fragen im Bereich der öffentlichen Bibliotheken. Berlin: Deutscher Büchereiverband Arbeitsstelle für das Bücherwesen, 1972. 37p. (Bibliothekdienst, Beiheft 72) (in German) (international, Germany)

The author presents a legal treatise concerning the scope of the copyright law in the Federal Republic of Germany, particularly as it applies to the protection of copyrighted works vis-à-vis photocopying practices in public libraries. The questions of author consent, remuneration, responsibility, and third-party copying are discussed.

Doyle, Francis R. A nickel for your ----: the copyright problem. Law Library Journal, 69(2):203-9; May, 1976.

The author discusses the copyright problem as it affects libraries, particularly with regard to photocopying practices. He deals with both the economics problems of libraries and their advocation of an educational use exemption from copyright litigation. This is followed by paragraphs on infringement, fair use, with mention of the Williams & Wilkins case, and compensation to copyright owners (in the form of taxation, publishers' contracts and litigation). The author concludes with a kind of "pox on both your houses" summation suggesting that libraries have no one but themselves to blame in this predicament because the law is "fairly explicit", and that "submitting to the copy machine was an error in judgment"; additionally, the copyright owners have failed to exercise the rights granted under the present laws and "have not sought other alternatives to compensation." Both parties are capable and should be able to meet the demands of the reprographic revolution.

-E-

Ehrenberg, Ralph. Reproducing maps in libraries and archives: the custodian's point of view. Special Libraries, 64(1): 18, 20-24; January, 1973.

The author outlines three problems of map reproduction from the custodian's point of view: 1) map accessibility; 2) photographic reproduction (technical aspects, not legal); 3) map handling. (2) and (3) are technical issues and outside the scope of this report. The first, accessibility, is controlled largely by factors outside the custodian's control, such as copyright law, agency and donor restrictions, and national security classifications. The author discusses the "fair use" concept with respect to photocopying copyrighted maps, the position of the Joint Libraries Committee on Fair Use in Photocopying, and the impact of the Williams & Wilkins case on photocopying practices.

Ellsworth, Ralph E. Library information machines not to undercut authors. Authors' Guild Bulletin, 5-6; October-November, 1971. (news)

E.P.C. and B.D.C report on 1974-75. The Bookseller, no. 3618: 2248-52; April 26, 1975. (international, Great Britain)

The Educational Publishers Council and the Book Development Council, two divisions of the Publishers Association in Great Britain have extracts from their annual reports discussed. Mention is made of copyright law revision and its impact on copying of copyrighted material in schools.

Esezobor, J.E. Copyright implications for reprographic services in libraries. UNESCO Bulletin for Libraries, 25(1): 18-26; January-February, 1971. (international)

The author surveys laws governing photographic reproduction by libraries and other institutions in various countries as well as current practices followed by libraries in photocopying in these countries. He states that the main problems are "those raised by technological development in book reproduction media, and the ambiguity of the terms used by the statutes and courts in their attempts to define certain limitations placed on the exclusive right of the author." The law with respect to developing countries is also briefly discussed. He concludes that unless libraries can support reprographic services they cannot effectively support research, and that reprography may augment copyright if both have as the same goal the spread of knowledge.

-F-

Ferguson, R.C. Plagiarism of unpublished works: is there an inference of access? Journal of the Beverly Hills Bar Association, 8: 27-36; September, 1974.

Finkelstein, Herman. Copyright problems on campus. The College Counsel, 6(1): 203-19; 1971.

Finkelstein, Herman. Copyright problems on campus. ASCAP Today, 6(2): 24-25; January, 1974. (news)

Fotokopieren in den USA--Schutzlosigkeit von Autoren und Verlegern? Gewerblicher Rechtsschutz und Urheberrecht, Int. Teil, no. 6: 246-51; June, 1974. (Williams & Wilkins)(in German)

Frankfurt, 1974, by a special correspondent. The Bookseller, no. 3593: 2488-92; November 12, 1974. (international)(news)

Freitag, B.J. It's time that teachers be heard on pending copyright legislation. Today's Education, 64: 65; September, 1975. (news)

Freitas, Denis de. Reproductions for private use, and for educational and library purposes. Revue International du Droit d'Auteur, no. 71: 30-57; January, 1972. (international, United Kingdom)

The author begins by reviewing a portion of the Berne Copyright Convention and noting the almost total lack of unanimity among the member countries as to the scope of authors' rights to reproduction. He then discusses the right of an individual to reproduce a copyrighted work for private purposes with respect to various European copyright laws and offers a critique of the United Kingdom's "fair dealing" clause. The subject then turns to the area of reproductions for educational and library uses, and the difficulties connected with obtaining individual permissions under the present United Kingdom law are reviewed as well as the various solutions that have been offered in other countries. The author concludes by proposing that: 1) the present English law extending control by copyright owners to reproductions for private use should be abolished; 2) a fund should be established out of which payments could be made to owners in respect of private reproductions; and 3) a compulsory licensing system should be established for licensing reproductions for educational use and for libraries.

Frotz, Gerhard. Zur Reform Vervielfältigung zum eigenen Gebrauch. Osterreichische Blätter für Gewerblichen Rechtsschutz und Urheberrecht, 21(5): 105-12; September-October, 1972. (in German)
(international, Austria)

Questions concerning the adequacy of the present Austrian copyright law for protection of authors and performing artists are discussed. The slant of the article is toward a consideration of the rapid technological advances in all forms of reproduction of copyrighted material, but limited to reproductions for personal use. The author raises the issue as to whether, in light of these advances, Austria should consider revision of the present law, and offers several suggestions toward this end.

Fry, Bernard M. & White, Herbert S. Economics and interaction of the publisher-library relationship in the production and use of scholarly and research journals. Final Report, Grant No. GN-41398. Washington, D.C.: U.S. National Science Foundation, Office of Science Information Service, 1975. 401p.

An extensive survey was undertaken to determine publisher-library interactions. Statistics are broken down into those covering various strata of libraries and publishers on the basis of type, size, function, locale, etc. Detailed presentations of the sampling techniques are made as to the patterns of photocopying and distribution of journals. The section on specific findings (pp. 58-162) are expecially noteworthy, as is the chapter on alternative publication systems. The book is an invaluable aid to any researcher in this sensitive area.

Garfield, Eugene. Copyright revision bill requires priority action to meet needs of the information revolution. Current Contents, no. 16: 5-8; April 18, 1973. (news)

Gehlin, Jan. Le contrat suédois relatif aux reproductions effectuées dans les établissements scolaires. Interauteurs, no. 184: 92-97; 1973. (in French) (international, Sweden)

This report analyzes the contract concluded on March 5, 1973 between the Swedish government and organizations representing all categories of writers, regarding payment of remuneration to the authors for reproduction of their works. The scope of the contract is at the scholastic level - corresponding to primary and secondary school establishments, which are considered in a different light than institutions of higher education, such as universities and professional schools.

Gerbrandy, S. The Netherlands solution to the problem of reprography. Copyright, 11(2): 47-51; February, 1975. (international, Netherlands)

The article is a concise review of the facets of the Netherlands Copyright Act amended January 7, 1973 regarding the photocopying provisions in Articles 16b and 17. Briefly stated, there are four rules: a) reproduction of "small" portions of books or the complete text of out of print books or of articles in periodicals is permitted for private use; b) public service institutions (libraries and educational institutions) have special consideration; c) enterprises may make or order copies as in (a) provided their use is restricted to employees of the enterprise and the number of copies does not exceed reasonable needs; d) reproductions may not be transferred to third parties. In all cases, remuneration must be paid to the author of the original work at the rate of 0.10 guilder per copied page or 0.025 guilder per copied page for copies made by primary and secondary educational institutions. Each of the four provisions is discussed with reference to the scope and effect of the law on present copying practices.

Goldstein, Paul. Private consumption of public goods: a comment on Williams & Wilkins Co. v. U.S. (487 F 2d 1345). Bulletin of the Copyright Society, 21(3): 204-13; February, 1974. (Williams & Wilkins)

Grannis, Chandler B. Copyright protects the public interests. Publishers' Weekly, 202(21): 48; November 20, 1972. (Williams & Wilkins) (news)

-H-

Haas, W.J., et al. Research libraries group reply to the publishing world regarding the effect such consortia will have on book and serial sales. College and Research Libraries News, 9: 213-5; October, 1974. (news)

Hagel, Raymond W. Some outspoken opinions from Raymond Hagel, Chairman of MacMillan. Publishers' Weekly, 206(4): 36-41; July 22, 1974.

An interview with the president and chairman of the board of MacMillan, Inc., chiefly concerned with publishing operations is presented. Of relevance here is Mr. Hagel's strong advocacy of publisher support of Title II in S. 1361 concerning creation of a National Commission on New Technological Uses of Copyrighted Works. Although he feels the pending Williams & Wilkins case will have great impact on the copyright controversy, he also feels that putting off support for Title II "in the interval is foolhardy."

Hamilton, L. Clark. Copyright, technology and compulsory licensing. Inter-Media, 2(6): 12-14; May, 1975. (news)

Hamilton, L. Clark. Technology and the law in copyright. Idea, 17(3): 25-34; Fall, 1975.

The impact of the new technology is reviewed as it affects United States and international copyright law. Particularly emphasized are the areas of reprographic reproduction, computer, and photocopying in libraries. As many instances are noted in the literature on the proposals for compulsory licensing systems as solutions to these problems, the author discusses some of the potential problems that might arise in the administration of such systems.

Hart, Bruce. Copyrights: concurrence, revision, and photocopying - Williams & Wilkins Company v. the United States (487 F 2d 1345): a "holding operation" by the United States Court of Claims. Dickinson Law Review, 79(2): 260-308; Winter, 1975. (Williams & Wilkins)

Harvell, Michael. Copyright--fair use--government photocopying does not infringe journal copyright. Boston University Law Review, 54: 689-701; May, 1974. (Williams & Wilkins)

Harvey, T. Problems of plagiarism. Times Educational Supplement, 3095: 68-69; September 20, 1974. (news)
(international, Great Britain)

Harvey, William. Orphans of the storm. Scholarly Publishing, 1(3): 229-305; April, 1970.

The author reviews the financial state of affairs in publishing with respect to reprints, especially in the area of university publishing. University publishers wrongly assume that unit costs of reprints will be half or less than publishing of the original, and that reprints are sure things, yet more and more scholarly books are going out of print every day. Reprint publishers favor the user, ignoring interests of author and publisher. Fair use muddies up the situation even more. He suggests that scholarly publishers take up reproducing their own out-of-print stock themselves, making titles available in a copyflo or equivalent form, and not dropping the titles from the original Publishers' Trade List Annual, which would have the effect of the title's remaining in Books in Print. Pricing would still be controlled by the publisher, the author would still receive royalties, and no lapse in availability would occur.

Heilprin, Laurence B. An experimental copyright moratorium: study of a proposed solution to the copyright photocopying problem. Final report to the American Society for Testing and Materials (ASTM). Washington, D.C.: Committee to Investigate Copyright Problems Affecting Communications in Science and Education, Inc., July, 1970. 60p.

A non-profit, voluntary system of access, permission, and payments to be called a "copyright clearinghouse" is proposed by the Committee to Investigate

Copyright Problems (CICP) joined by the American Society for Testing and Materials. A preliminary experiment involving client organizations of the ASTM guaranteed against infringement in making up to fifty copies of any ASTM publication while allowing analysis of the copying to be carried out by CICP. Unfortunately, government support for the CICP ended and the amount of data obtainable and analyzable was greatly reduced. The report based on the available data states that the clearinghouse is feasible and would resolve the copyright photocopying problem.

Henry, Nicholas L. Educators and the politics or neopublishing. Educational Administration Quarterly, 7(2): 60-73; Spring, 1971.

The author reviews the development of photocopying and its role in education, and traces the efforts of educational and university library associations in the present revision of the United States Copyright Law now before Congress. He concludes by examining the relationship between neopublishing and the information explosion.

Henry, Nicholas L. Copyright, public policy, and information technology. Science, 183(4123): 384-91; February 1, 1974.

The author discusses two "information technologies" - computer-based information storage and retrieval systems and photocopying techniques - which are the causes of major controversies in copyright law interpretations. In the section dealing with photocopying, publishing and research, the author reviews the upwardly spiraling numbers of photocopies being made; the surveys which have been undertaken on photocopying as related to page costs, user groups, attitudes of those in the publishing industry and those research; and finally some of the solutions being offered. He suggests that the proposed National Commission on New Technological Uses of Copyrighted Works might be a possible agency for arriving at workable agreements between copyright owners and users and lists seven guidelines for developing an information policy.

Henry, Nicholas L. Copyright: its adequacy in technological societies. Science, 186(4168): 993-1004; December 13, 1974.

The "utility of copyright as a public policy for knowledge management in a society that possesses a high level of information technology" is analyzed from the development and extent of copyright protection to its failures in the areas of public domain policy, censorship, "for-profit" limitations, and lastly, a lengthy discussion of the necessities for statutory inclusion of a "fair use" dimension in copyright law. Several proposed modifications and alternatives are reviewed, including abbreviated copyright duration, non-exclusive licensing, media-tailored copyright, computerized accounting systems, group purchasing, page charges, and extension of copyright to personal behavior. He suggests neither an extension nor retraction in the scope of copyright until the relations between copyright and neopublishing are fully explored.

Hill, H.E., il. Copyright revisited: fair use. AV Instrument, 19: 35-36; June, 1974. (news)

Holbrook, Lanny R. Copyright infringement and fair use. University of Cincinnati Law Review, 40(3): 534-47; Fall, 1971.

The article is a discussion of the elements contributing to the determination of best interests. Who should prevail - the interests of society or the author or creator of original works? The body of the article is devoted to a thorough discussion of the common law (as opposed to statute) of fair use concerning its origins, developments, scope, and an "attempt to delineate its permissible boundaries", as a consequence of its use as a defense in infringement cases. Numerous case studies are brought up in the discussion.

Hollingsworth, Robert J. Copyright--libraries--unauthorized library photocopying of articles from copyrighted medical journals constitutes copyright infringement. University of Cincinnati Law Review, 41(2): 511-19; 1972. (Williams & Wilkins)

Holroyd, Michael. The 'system' in manuscript libraries, based on prestige and the exclusivity of material aggravates...the wrongs of copyright. Library Journal, 101(9): 1081-83; May 1, 1976. (news)

Holusha, John. If Xerox had pushed its process. Washington Star-News, A-3, columns 1-2 (night final edition); April 16, 1974. (news)

Horowitz, Israel. Photocopying spurs 'pubs' new 'fair use' standards. Billboard, 87(35): 1,14,50; August 30, 1975. (news)

Horowitz, Israel. Fight illegal duplication; MPA pushing fair use of music plea. Billboard, 87(51): 3; December 20, 1975. (news)

House subcommittee holds hearings on copyright bill. BNA's Patent, Trademark & Copyright Journal, no. 228: A6-A13; May 15, 1975. (news)

-I-

IFLA/IALL Working Group on Copyright, Julius J. Marke, Chmn. Copyright (preparatory draft). May 23, 1974. 22 type-written pages. (Chapters 4 (revised) and 5 from the Working Document for the November 19, 1974 meeting of the group in Washington, D.C. in IFLA Journal, 1(1): 47-53; 1975). (international)

The purpose of the group was to present basic premises which should be considered in identifying copyright problems affecting libraries and information centers world-wide. The introductory sections present background issues affecting the present state of affairs and acting as frames of reference for the committee. The meat of the paper begins with Chapter 3 which reviews the impact of the growth of copying practices on library services, research, documentation centers, educational institutions and publishing economics. Chapter 4 discusses the premises to be considered: a) two means of reducing the amount of photocopying, reprint services offered by publishers and the imposition of fees or licensing systems for photocopying privileges are discussed and the reasons why neither is acceptable presented; b) the position is taken that in view of the unacceptability of the alternatives in (a), librarians whould consider that non-profit copying for education and research is within the realm of "fair use"; c) no position should be taken on copyright and computerized storage and retrieval systems; d) no position with reference to developing nations; e) public lending right to publishers should be rejected; and f) recommendations of the committee should be drafted as an international instrument for guidance to nations concerned with these problems.

Inner London Education Authority. More evidence on copyright. The Bookseller, no. 3606: 360-67; February 1, 1975. (international, Great Britain)

Intergovernmental Copyright Committee. 11th session, Geneva, November, 1971. Report. Paris: Intergovernmental Copyright Committee; January 14, 1972. 19p. (international)

This is a report of the meeting of the committee which was established as part of the Universal Copyright Convention. Among other items on the agenda were the revision of the Universal Copyright Convention (July, 1971), the necessity of resolving the photocopying problem with respect to copyrighted works, the problems resulting from computerized storage and retrieval of these works, and a report of recommendations adopted by international organizations in this field.

Intergovernmental Copyright Committee. 12th session, Paris, December 5-11, 1973. Report. UNESCO Copyright Bulletin, 8(1): 18-41; January, 1974 (Also appears in Copyright, 10(2): 52-56; February, 1974). (international)

Of particular interest in this report are the paragraphs (nos. 14-59) on the subject of reprographic reproduction of works protected by copyright. Discussion is presented by participants from member states on this issue without conclusion. However, in consideration of the importance of the problem, even though it felt that now was not the time for a definitive stand, the committee voted to continue the study by means of appointing a subcommittee for this purpose.

International Copyright Union. 6th session (2d estraordinary), Paris, December 5-11, 1973. Report. UNESCO Copyright Bulletin, 8(1): 18-41; January, 1974 (Also appears in Copyright, 10(2): 32-39; February, 1974). (international)

See preceding entry for summary. The two separate entries in Copyright are presented as a single entry in the UNESCO Copyright Bulletin.

Jeanneret, Marsh. Copyright and the scholar. Scholarly Publishing, 1(3): 227-28; April, 1970. (news)

Jeanneret, Marsh. Information retrieval and the decision to publish. Scholarly Publishing, 1(3): 229-43; April, 1970.

The author reviews the technological evolution of library science into modern, computerized information storage and retrieval systems, and the threat posed to authors in terms of creation of new works and publishers in terms of decisions as to whether or not to publish. Computerized ISAR systems complement library lending services and the author states that "there can be no escaping the conclusion that a free computer input privilege would inhibit publishing, and thus creative writing as well." He suggests a compulsory licensing provision for libraries who wish to store verbatim copyrighted material in a central retrieval system in return for a fee paid to the copyright owner, and discusses methods of establishing and reviewing equitable remuneration.

Jovanovich, William. The Universal Xerox Compiler Machine. The American Scholar, 40(2): 249-55; Spring, 1971.

Justices to decide if libraries must pay copyright for photocopying items. Wall Street Journal, 183(104):7, cols. 3-4; May 29, 1974. (news)

-K-

Kaplan, Benjamin & Brown, Ralph S., Jr. Cases on copyright, unfair competition and other topics bearing on the protection of literary, musical, and artistic works, 2d ed. by Brown. Mineola, N.Y.: Foundation Press, 1974. 997p.

The second edition of a 1960 work which recognizes the increased role of photocopying, computer uses of copyright material, and also cable television is presented. The format is generally a presentation of a particular topic followed by several cases illustrative of the point. Of particular interest is the section on Fair Use in Topic E, Infringement. Several cases are mentioned, including Williams & Wilkins (before trial commissioner Davis) with an almost complete transcription of his written decision for the plaintiff, and a series of notes and questions following.

Katzenberger, Paul. Urheberrecht und Dokumentation: Abstracts - Fotokopien - elektronische Datenbanken. Gewerblicher Rechtsschutz und Urheberrechts, 75(12): 629-40; December, 1973. (international, Germany) (in German)

The author studies the relationship between German copyright law (Federal Republic of Germany) and various forms of documentation - abstracting/indexing, photocopying, and data banks. Because of the rapid increase in computer and reprographic technology, he urges a revision in the fair use provisions of the German law. A section on the reprographic situation with respect to copyright in other countries is also included.

Keenan, Stella and Stillman, Mary E., issue ed. Information Retrieval Colloquium, 9th. Philadelphia, Pa., May, 1972. Panel on copyright. The copyright controversy: issues and opinions. Drexel Library Quarterly, 8(4): 377-602; October, 1972. (Williams & Wilkins)

Kent, Allan and Lancour, Harold, ed. Copyright: current viewpoints on history, law, legislation. New York: Bowker, 1972. 125p.

The book presents in each of its chapters representative opinions from the point of view of librarians, publishers, and computer scientists and authors, in addition to a brief history and a factual analysis of United States and international copyright conventions. Especially pertinent are the chapters on copyright as it affects libraries, with emphasis on reproduction and fair use, photocopying, and the not-for-profit publishers, and the library user with respect to copying and fair use.

Kerever, André. La reprographie et les normes internationales en matière de droit d'auteur. Il Diritto di Autore, 46(3): 330-40; July-September, 1975. (in French)
(Reprinted in Copyright, 12(7/8): 188-96; July-August, 1976) (international)

Kim, Chin. Librarians and copyright legislation: the historical background. American Libraries, 2(6): 615-22; June, 1971.

This is a brief review of the history of copyright in the United States and the role of librarians and the American Library Association in the enactment of the 1891 and 1909 acts, and their current efforts toward revision of the present act. Primary concern is over the status of the "fair use" doctrine with respect to library photocopying, and discussion is presented on the differing versions of House and Senate bills on this matter. Two questions are posed: 1) are librarians to continue to insist on endorsement of the "not-for-profit" principle in the final version of the bill? and 2) how are librarians to appraise note-taking of users in view of cheaper and easier means of photocopying?

Koch, William H. Copyrighting physics journals. Physics Today, 27: 23-28; February, 1974.

Koch, William H. Copyright in the next ten years: views of a science publisher. New York: American Institute of Physics, October 27, 1975. 9p.

The text of a report presented at a recent American Society for Information Science meeting stating occasions when scientific articles from American Institute of Physics journals have been copied without permission, concludes that increased growth of computerized networks will greatly accelerate the financial crises in the publications industry.

Kolle, Gert. Reprographie und Urheberrecht; ein rechtsvergleichender Beitrag zur Rolle des Urheberrechts in der Informationsgesellschaft. Gewerblicher Rechtsschutz und Urheberrecht, Int. Teil, no. 6: 201-14; June, 1975. (international)(in German)

This is an elaboration of a topic originally presented as a talk before the International Congress for Reprography and Information on April 16, 1975, which addressed the topic of the role of copyright in the information society, and specifically the interactions of reprography and copyright.

Kolle, Gert; Katzenberger, Paul; & Peter, Franz-Wilhelm. Reprographische Vervielfältigung und Urheberrecht; Arbeitsergebnisse und Reformvorschläge des Fachausschusses für Urheber- und Verlagsrecht und seines Unterausschusses "Reprographie". Gewerblicher Rechtsschutz und Urheberrecht, 77(9): 468-82; September, 1975. (in German) (international, Germany)

This is a report on the results and recommendations of the Committee of Experts on Copyright and publishing Law of the German Society for Industrial Property Protection and Copyright. The agenda was for revision of existing Federal Republic of Germany Copyright Law regarding the provisions covering reprography, and the report is based on the analysis performed by, and proposals of, the Reprography Subcommittee.

Kolle, Gert and Ulmer, Eugen. Einspeicherung geschutzer Werke in automatische Informations- und Dokumentationssysteme. Gewerblicher Rechtsschutz und Urheberrecht, Int. Teil, no. 3: 108-16; March, 1976. (international)(in German)

-L-

Lacy, Dan. A statement on H.R. 2223: the general copyright revision bill. Publishers' Weekly, 208(21): 25-28; November 24, 1975. (news)

Lahore, James. Photocopying in Australian libraries. Developments in copyright law. International Journal of Law Libraries, 4(1): 32-37; March, 1976. (international, Australia)

This is an analysis of the growing dispute between the Australian Copyright Council and Australian universities over the "largely uncontrolled use of photocopying machines in libraries, particularly University libraries", which culminated in the recent suit, University of New South Wales v Moorhouse and Angus & Robertson (Publishers) Pty Ltd. The Council in attempting to achieve a judicial determination of whether or not universities were liable for photocopying done on unsupervised machines deliberately arranged to have two copies of a story by Frank Moorhouse (approximately ten pages) contained in a 220-page collection of twenty short stories photocopied. The author was unaware of this until after the fact. The Supreme Court of New South Wales, while granting that infringement occurred, held that the University was not liable, but this decision was overturned by the High Court. The ramifications of this decision as it affects both universities and publishers is further discussed.

Legality of photocopying copyrighted publications. Williams & Wilkins Co. vs. the United States. William & Mary Law Review, 13(4): 940-48; Summer, 1972. (Williams & Wilkins)

Libraries and copyright: a summary of the arguments for library photocopying. Chicago, Ill.: American Library Association, June, 1974. 51p.

The booklet presents a summary of the arguments advanced in favor of library photocopying and a status of these arguments with regard to current legislation and court decisions. An analysis of the Williams & Wilkins case follows, including a presentation of statements by the American Library Association, the Association of Research Libraries, and the National Library of Medicine, and also individuals in the library community. The report concludes with congressional decisions on copyright and the pertinent sections (107 and 108) on photocopying in the revised bill, S. 1361.

Library associations speak out on copyright. Library Journal, 100: 1267; July 1, 1975. (news)

Limperg, Th. De problemen van de reprograjische reproduktie van auteursrechtelijk beschermde geschriften. Bijblad bij De Industriele Eigendom, 41(10): 163-71; October 15, 1973. (international, in Dutch)

See Limperg, Th. below for the abstract of the French/English version of this article.

Limperg, Th. Les problèmes de la reprographie des oeuvres imprimées protegées par le droit d'auteur. Revue internationale du Droit d'Auteur, no. 78: 121-91; October, 1973. (international)(French/English)

This is an extensive review of the photocopying problem beginning with discussion on terminology and methods of reproduction, followed by examples of various national legislations occurring in response to this growing area; the risks of the technology to authors and publishers as regards potential loss in sales of copyrighted works; a brief history of national and international resolutions to this effect; a survey of the existing limitations on the right of reproduction (with special emphasis on the "fair use" concept); reproduction by private individuals, by libraries and documentation centers, by scientific institutions, by commercial, for-profit organizations, and others; and concludes with a discussion on the practical consequences of existing practices, stressing the need for international regulation and relatively uniform national legislations "to check the harmful effects of reprography while there is still time."

Line, Maurice B. & Wood, D.W. Effect of a large-scale photocopying service on journal sales. Journal of Documentation, 31(4): 234-45; December, 1975.

A survey of the requests for serials to the British Library Lending Division was conducted in order to collect evidence on the claims that journal subscriptions are either declining or not increasing proportionally, that photocopying has grown enormously, and that this growth is directly impacting journal sales. Results indicated that although journal prices are higher, university budgets are devoting greater portions to subscriptions, that most requests for loans are on widely-held and circulated core journals, and that photocopying of in-print journals is small.

Line, Maurice B. & Wood, D.W. Photocopying and journal sales: a reply. Journal of Documentation, 32(3): 204-6; September, 1976.

Reply to Van Tongeren, 1975 below on original paper above.

Lingeman, R.R. Copyright and the right to copy. New York Times Book Review, p. 63, November 17, and p. 47, November 24, 1974. -- Comment by J.E. Skipper (letter) with rejoinder, the Book Review, p. 34-35, January 12, 1975. (news)

Long, Janice R. Copyright bill gets workover from all sides; educators, librarians, others air views on library photocopying provisions of copyright bill in House subcommittee hearings. Chemical & Engineering News, 53(21): 14-15; May 26, 1975. (news)

Lottman, Herbert R. Frankfurt 1974: an inflationary time, prudence, caution and high professionalism. Publishers' Weekly, 206(19): 46-53; November 4, 1974. (international)

A report is given on the twenty-sixth Frankfurt Bookfair covering topics ranging from East-West copyright relations conferences with the Soviet Union to a separate section dealing with international attitudes among publishers of scientific and technical journals toward the increased incidence of unauthorized photocopying.

Lottman, Herbert R. Photocopying: how is Europe handling the problem? _Publishers' Weekly_, 206(22): 26-27; November 25, 1974. (news) (international)

Lottman, Herbert R. Copyright erosion worries some foreign publishers. _Publishers' Weekly_, 207(22): 14; June 2, 1975. (news) (international)

Lovallo, C.G. Fair is fair: solution to the copyright problem: faculty development of instructional materials. _College Management_, 9: 4; October, 1974. (news)

Low, Edmon. _Copyright and fair use practices_. Phonotape. Los Angeles: CREDR Corp., 1973. 1 cassette.

Low, Edmon. What Edmon Low told the House subcommittee about libraries. _American Libraries_, 6(7): 410-11; July-August, 1975. (news)

Lukac, George, ed. _Copyright--the librarian and the law; proceedings of the eighth annual symposium sponsored by alumni and faculty of the Rutgers University Graduate School of Library Service_. New Brunswick, N.J.: Bureau of Library and Information Science Research, Rutgers University Graduate School of Library Service, 1972.

This book is divided into four parts: 1) introductory material; 2) five articles dealing with various viewpoints on copyright with respect to fair use and photocopying; 3) a panel discussion among the five authors from part two; and 4) appendices, including copies of the present copyright law, changes in Copyright Office Regulations, and the copyright revision bill. The individual authors of the five articles are Barbara Ringer on present copyright law, Charles H. Lieb on a publisher's view of photocopying, Irwin Karp on an author's view of photocopying and copyright, Paul G. Zurkowski on proprietary rights, and Verner W. Clapp on copyright legislation.

Lukac, George. The copyright controversy. _Writer's Yearbook_, 84-87; 1973.

This is a summary of the symposium sponsored by Rutgers University Graduate School of Library Service (see Lukac, 1972). At the center of the controversy is the effort to revise the existing copyright law, and the author states two factors inhibiting the process: 1) introduction and improvement of new technological devices - photocopiers, computers and cable TV in particular - which negate newly-drafted laws as quickly as they are written; and 2) the tacit position that Congress has taken not to pass a law until authors, publishers and librarians have reached an accord. A brief review of the present law and the new proposals causing the greatest furor, extension of the term of protection, fair use, and change in common law protection, is presented, followed by viewpoints from each of the protagonists. There is agreement among authors and publishers that it is not photocopying per se that is the problem, but the lack of remuneration from such copying. From all points, the dilemma of the scope of fair use is still the major hurdle preventing accord.

-M-

MacKenzie, John P. Supreme Court faces issue of copyright vs. copying. The Washington Post, p. A2, columns 1-3; December 18, 1974.
(Williams & Wilkins)(news)

Maclean, Ann W. Educational and copyright law: an analysis of the amended copyright revision bill and proposals for statutory licensing and a clearinghouse system. Virginia Law Review, 56(4): 664-89; May, 1970. (Later published in Copyright Law Symposium (ASCAP), 20(1): 1-35; 1972)

The author analyzes the work in progress on revision of the present copyright bill with respect to the impacts on the field of education. The doctrine of fair use is greatly discussed in regard to what teachers can and cannot do, and protection of teachers who do infringe. Also covered is the problem of the relationship of the copyright holder's right to control the uses to which his material is put and the educational uses of protected material. She proposes a system of compulsory licensing and a clearinghouse system for educational institutions in terms of; 1) filing procedures for obtaining a license; 2) fee-setting for licenses; and 3) use of a clearing house system for distribution of license fees to copyright owners.

Marke, Julius J. Copyright law and photocopying in libraries. New York Law Journal, 172(11): p. 6, cols. 1-7; December 10, 1974.

This is an article dealing with the confrontation over library photocopying of copyrighted materials between consumers of intellectual property and the copyright owners, and the effects of two recent developments, the Williams & Wilkins case and the forthcoming revision of the copyright law, on this battle. The author feels that section 108 of the revised copyright bill is so restrictive on the scope of library photocopying that network planning and the possibility of a national library and information system will be virtually destroyed. Further points from both sides in connection with fair use, Williams & Wilkins, publishing economics, licensing, and the administrative costs of licensing are discussed.

Marke, Julius J. Photocopying and the copyright law in the United States. International Journal of Law Libraries, 3(2): 169-70; 1975. (news)

Maslansky, Harris. Copyright: what copying is right? Science Teacher, 40(5): 49-50; May, 1973. (news)

McCormick, Marilyn G., comp. The Williams & Wilkins Case: The Williams & Wilkins Co. v. the United States. Volume I. New York: Science Associates/International, 1974. 275p. (Williams & Wilkins)

Melichar, Ferdinand. Photocopying of works protected by copyright. GEMA News, no. 14: 51-60; December, 1975. (in English, French & Spanish) (international)

Recent developments in copyright law for several European countries and the United States are reviewed with respect to the impact of photocopying and its relationship to publishing economics. A possible solution is the suggested levying of a surcharge on individual copying machines rather than on material to be copied; national uathors' societies would handle distribution of royalties.

Mitchell, Jerry C. & Duvall, Charles R. A study comparing the opinions of music publishers and music educators concerning copying of extra parts. July, 1974. 94p.

The report of a study on the opinions of music publishers concerning music educators making single copies of individual parts of purchased arrangements is presented. The study covers approximately fifty each of publishers, band directors and choral directors. Results showed that: 1) the law is clear and that there is no reason for educators to make unauthorized copies of parts; 2) however, educators are making copies and publishers are aware of this, but have as yet taken no legal action; 3) variations from publisher to publisher in standard instrumentation dilute the value of standard instrumentation ordering; 4) publishers and educators disagree as to the seriousness of the copying problem.

MLA argues fair use interpretation of copyright law. Virginia Librarian, 20: 18; December, 1974. (news)

Mueller, Jöran. Kopierung för undervisningsändamål i Sverige; redogörelse för ett nyligen träffat avtal. NIR, no. 4: 384-96; 1973. (in Swedish) (international, Sweden)
(See Mueller, Jöran below for abstract of English language version of this article)

Mueller, Jöran. Photocopying in Swedish schools. Stockholm, Sweden: Ministry of Education, August 31, 1973. 10p.
Reprinted by the Council of Europe, Council for Cultural Cooperation, Strasbourg, France (DECS/INF (73)9). (international, Sweden)
(Later published in Prospects, 4(2): 263-69; Summer, 1974)

The provisions for photocopying for educational purposes are now contained (as of July 1, 1973) in a special agreement between the Swedish government and all categories of writers' organizations, not under statute law. Negotiations for this agreement began in 1967 under the theory that the legal support given in the Swedish Copyright Act and Photography Act was considered insufficient, and that wide-scale copying was occurring on a large basis without the knowledge or consent of the copyright owner. This agreement extends at this time only to primary and secondary schools; negotiations are still going on with respect to copying in universities.

Müller, Winfried. Problems of copyright arising from microfilming. <u>Informatik</u>, 21(6): 42-43; 1974. (news)

Murdock, Kent. Williams & Wilkins Co. v. U.S. (487 F 2d 1345): library photocopying of copyrighted materials. <u>Utah Law Review</u>, 1974: 127-44; Spring, 1974. (Williams & Wilkins)

Murphy, James P. The legal effects of Williams & Wilkins. <u>Publishers' Weekly</u>, 202(24): 11-12; December 11, 1972. (Williams & Wilkins)(news)

Nasri, William Z. <u>Impact of reprography on the commercial publication of scientific and technical periodicals with particular reference to the case of "The Williams & Wilkins Co. vs. the United States"</u>. Ph.D. dissertation. Pittsburgh, Pa.: University of Pittsburgh, 1975. (Williams & Wilkins)

Nasri, William Z. <u>Crisis in copyright</u>. New York: Marcel Dekker, 1976. 174p.

The thesis upon which the subject matter of this book is based is the crisis currently racking the world of information - the impact of reprography upon producers and users of copyrighted material. Specifically, the question is asked as to whether commercial publishers of scientific and technical periodicals are suffering losses in subscriptions because of increased photocopying by libraries, researchers, students, etc. The first three chapters present background material on copying trends over the years, and review the present (1909) law. Chapter 4 is an analysis of the Williams & Wilkins case. Based on the publishers' claims in this case, a survey was conducted (see also Nasri, 1975) to determine the actual number of subscriptions by members of professional societies. As no cooperation was received from publishers in this effort, the original scale of the survey was reduced and focused instead on the four journals which also happened to be the four subjects of litigation in the Williams & Wilkins case. The evidence of the survey indicated that reprography had no significant effect on subscriptions and was not the primary reason for loss of subscriptions.

National Council for Educational Technology. <u>Copyright and education: a guide to the use of copyright material in educational institutions. Working Paper no. 8.</u> London: Councils and Education Press, Ltd., 1972. 91p. (international, Great Britain)

This report deals with the copyright laws of Great Britain as they affect teachers regarding use of copyrighted materials. It covers not only photocopying of printed materials, but also performances, copies of films, tapes, etc. Each area of consideration is followed by a statement concerning the scope of the law and conditions for use of the materials under the law. Recognizing the difficulties in interpretation for educational purposes, new approaches and possible courses of action in terms of agreements with publishing and performing arts associations or in development of central agencies are suggested for future consideration.

National Science Foundation, Office of Science Information Service. U.S.-U.S.S.R. copyright negotiations on scientific and technical journals. Information: News and Sources, 6: 203-5; September, 1974.
(international)(news)

Navasky, Victor S. In cold print: copyright and wrong. New York Times Book Review, p. 2; October 21, 1973.
(news)

Navasky, Victor S. In cold print: copy rights. New York Times Book Review, 80(29): 3,22; July 20, 1975.
(news)

Nelson, W.D. Library photocopying and copyright extension. Wilson Library Bulletin, 49: 333; January, 1975.
(news)

New rules on photocopy limits and classroom use set forth in full text of copyright addenda. American Libraries, 7(10): 610-11; November, 1976.
(news)

New coalition aims for copyright protection. Publishers' Weekly, 207(12): 15; March 24, 1975.
(news)

The New South Wales judgment. The Bookseller, no. 3585: 1732, 1734; September 7, 1974.
(international, Australia)
(news)

Nimmer, Melville B. 1972 supplement to cases and materials on copyright and other aspects of law pertaining to literary, musical and artistic works. St. Paul, Minn.: West Publishing, 1972. 110p.

The 1972 supplement to his earlier mammoth work updates the material presented there by inclusion of developments in copyright law throught June, 1972. With respect to photocopying, the most relevant sections are those on the recent developments in xeroxing with regard to copyright infringement and the scope of compulsory licensing provisions. The chapter and section headings are reproduced from the original supplement in this supplement.

Nimmer, Melville B. Photocopying and record piracy: of Dred Scott and Alice in Wonderland. UCLA Law Review, 22(5): 1052-65; June, 1975.

Nitecki, Joseph Z., comp. & ed. Directory of Library Reprographic Services, 5th ed. Weston, Conn.: Microform Review, Inc., 1973.

North, William. Williams & Wilkins. The great leap backward. American Libraries, 3(5): 528-30; May, 1972. (Williams & Wilkins)(news)

Nysonger, J. Stephen & Kennedy, Craig. Copyright: library use of photocopiers. Washburn Law Journal, 12(2): 235-40; Winter, 1973.
(Williams & Wilkins)

-O-

O'Donnell, Bernard. The teacher and the copyright law: response. Journal of Reading, 17(1): 15-16; October, 1973. (news)

Response to Caughran, 1973.

Oettinger, Elmer R. Copyright and the librarian. North Carolina Libraries, 31(4): 25-34; Conference issue, 1973. (Williams & Wilkins)

O'Grady, C. Copyright let up no gateway to freedom; Great Britain. Times Educational Supplement, 3137: 10; July 18, 1975. (news)
(international, Great Britain)

O'Neill, Judith D. Copyright - extensive photoduplication of copyrighted scientific journals does not constitute copyright infringement. University of Baltimore Law Review, 3(2): 306-17; Spring, 1974.
(Williams & Wilkins)

Pace, Eric. 4 libraries scored on copying plan. New York Times, at 27, col. 1-3 (city edition): April 12, 1974. (news)

Passano, William J. How photocopying pollutes sci-tech publication. Publishers' Weekly, 197(5): 63-64; February 2, 1970.
(Williams & Wilkins)(news)

Passano, William J. A statement from Williams & Wilkins to the editors, Publishers' Weekly, by William J. Passano, chairman of the board, Williams & Wilkins Co. Publishers' Weekly, 202(20): 17-18; November 13, 1972.
(Williams & Wilkins)(news)

Passano, William J. A publisher's view of photocopying. American Libraries, 5(5): 221; May, 1974) (Williams & Wilkins)(news)

Phillips, Jeremy. Copyright: towards a positive approach? Revue Internationale du Droit d'Auteur, no. 82: 34-65; October, 1974.
(international, Great Britain)

This is a lightly written article which begins with a philosophical essay on the British laws of copyright protection and then proceeds to touch on many individual "battleground(s)": the mode and duration of protection, copyright in computer programs and computer infringement of copyright, library lending rights, and photocopying. In the section on photocopying, the author takes no stance on "fair use" stating that there must be acceptance of the idea that photocopying is here to stay, and that means should be taken to provide a measure of compensation to copyright owners. Two proposals are reviewed - 1) a tax on the machine as is done with tape recorders in Germany, and 2) the formation of syndicates among periodical publishers to purchase shares in the reprography industry and distribute dividends from the shares proportionately to the rate of decline in the circulation of the respective periodicals.

Photocopying and fair dealing: submissions to the Departmental Committee. The Bookseller, no. 3585: 1728-32; September 7, 1974.
(international, Great Britain)

The Copyright Council of Great Britain was asked to submit its recommendations for dealing with the problem of photocopying copyrighted material to the Whitford Committee set up for the purpose of reviewing the 1956 Copyright Act. The Council endorses the idea of some sort of general licensing system as being "equally applicable to all categories of users." The Council supports the Publishers Association and feels that continued unimpeded copying will destroy the publishing industry.

Photocopying for researchers held legal. Chemical & Engineering News, 51(50): 12-13; December, 1973. (Williams & Wilkins)(news)

Photocopying: high court tie vote leaves issue to Congress. Science, 187: 934; March 14, 1975. (Williams & Wilkins)(news)

Pimsleur, Meira G., comp. Cumulative index, volumes 1-20, 1953-1973: Bulletin of the Copyright Society of the U.S.A. South Hackensack, N.J.: F.B. Rothman, 1975. 229p.

This is a very useful index divided into five sections. In each section the entry is followed by the number and pagination of the Bulletin for that entry. The first section is an alphabetical Author index, the second is an alphabetical table of all cases appearing in the Bulletin, the third a listing of legislative and administrative developments arranged alphabetically by country, and within each country, alphabetically by title of the act; the fourth section is an alphabetical list of conventions, treaties and proclamations; lastly, is a subject index with cross-references. Pagination in this section, if inclusive,denotes an article, and if a single page, refers to treatment of the subject on that page in some larger article.

Practicing Law Institute. Current developments in copyright law. Patent, Copyright, Trademarks, and Literary Property, Course Handbook Series, no. 31, 1972. 152p.

The book contains several articles on current issues of copyright law, including international copyright, misappropriation, and two articles of particular interest: Rights and licenses for use on microfilm, photocopying, computers, cassettes, and CATV by Bella L. Linden; and Computation of damages for copyright infringement by Bernard A. Helfat.

Publisher/library group urges photocopy fee plan. Library Journal, 100: 620; April 1, 1975. (news)

Publishers Association. Submission to the "Whitford Committee", appointed to consider "the Law on Copyright and Designs". London: The Association, 1974. (international, Great Britain)

Publishers under fire in 1974-75; P.A. reports on another eventful year. The Bookseller, no. 3616: 2084-92; April 12, 1975. (international, Great Britain)

The article summarizes the annual report of the Publishers Association in Great Britain dealing with, among other topics, copyright and education, photocopying, public lending right, and international copyright.

-R-

Rassudovskiy, V.A. Scientific information and copyright law. Arlington, Va.: Joint Publications Research Service (JPRS 52937), April 22, 1971. 13p.

Reitz, Norman E. Williams & Wilkins: (Williams & Wilkins Co. v. U.S. 172 USPQ 670): the impact of technology on copyright. Los Angeles Bar Bulletin, 48(12): 445-50, 468-72; October, 1973. (Williams & Wilkins)

Report of a survey of the copying of print materials by schools carried out by the Council for Educational Technology and the Publishers Association. London: The Association, January, 1975. 15p, 14 tables. (international, Great Britain)

Ringer, Barbara. Preliminary statement. Chapter III. Reproduction by libraries and archives. November-December, 1975 (unpublished)

Ringer, Barbara. Copyright in the 1980s--the Sixth Donald C. Bruce Memorial Lecture. Bulletin of the Copyright Society, 23(5): 299-310; June, 1976.

In an article dealing with the impact of technology on the future of authorship, the author expresses the view that copyright will remain "in the broad sense of the way a society chooses to treat its creative people." Exploring the United States copyright law over the past twenty-five years, somewhat ironically, she charts a recurring pattern of technological development, which is successfully exploited, followed by questions over possible infringement, discovery that the 1909 act contains no concrete answers, the bringing of a suit for copyright infringement, reluctant holding against the copyright owner by the courts, passing of responsibility to the Congress, and finally, compromise legislation which seems likely to be in the form of compulsory licensing. Without a good copyright law, she concludes that authors' independence will be destroyed.

Rosenfeld, H.N. Constitutional dimension of"fair use" in copyright law. Notre Dame Law Review, 50: 790-807; June, 1975.

Rosenfield, Harry M. Customary use as "fair use" in copyright law. Buffalo Law Review, 25(1): 119-40; Fall, 1975. (Williams & Wilkins)

-S-

Schulman, Michael. Unauthorized photocopying illegal, immoral: publishers avoid the villain role with schools, churches. The Music Scene, no. 279: 5, 17; September-October, 1974. (international, Canada)(news)

Selected bibliography on copyright (circular 2B). Washington, D.C.: U.S. Copyright Office, 1974. 2p.

Brief list, mostly historical references.

Senate passes copyright reform bill and proposes national commission on new technological uses of copyrighted works. Information: News and Sources, 6: 292; December, 1974. (news)

Shafter, Robert L. The photocopy industry and copyright: section 108 of the bill. Idea, 17(3): 25-42; Fall, 1975.

Sharp, Roy C. Licensing the photocopier. Scholarly Publishing, 1(3): 245-53; April, 1970.

The author reviews current status of copyright and speculates as to whether the advantages to society offered by copyright can be maintained in view of the technological advances. He discusses comparatively the present laws in the United States, the United Kingdom, and Canada with respect to photocopying and the interpretation of fair use, and states that most of the daily violations occur in libraries, universities, research and other educational institutions, who are prepared to pay from five to twenty-five cents per copy rather than go through time-consuming permission procedures. Two classes of proposals for fee-charging have been categorized: 1) per use, and 2) flat fee taxation of machines at purchase. Although the "per use" proposal has previously been regarded as impractical, the author believes that with the adoption of international book and per-iodical numbers, such a proposal is now feasible, and outlines propositions appli-cable to legislation in western countries for carrying out such a plan. Briefly, a collection agency would be established and assigned (voluntarily or compulsor-ily) photocopying rights, and collect and disburse royalty fees from institutions. Institutions would be provided with special copiers which would automatically re-cord ISBN or ISPN (ISSN) and the number of pages on paper or magnetic tape, and stamp each page of copy. A cartridge of the record would be sent periodically to the collection agency.

Sharp, Roy C. Copyright - shield or shroud. Copyright, 11(5): 113-18; May, 1975.

The impact of the logarithmically expanding use of copying processes engendered by the increased technology on both the copyright owners and users is discussed. Particular reference is made to librarian educators whose photocopying demands are seen as the most threatening to publishers. Lengthy rebuttal is offered to the two most frequently invoked claims under the "fair use" doctrine - single-copy reproduction of a journal article or portion of a book by or for an individual constitutes fair use, and specific exemptions should be made for research and educational purposes. The article also discusses three of the proposed solutions in the literature and makes a strong recommendation in favor of a blanket licensing system, requiring establishment of a royalty collection agency.

Shaw, Ralph R. Williams & Wilkins v. the U.S.; a review of the Commissioner's report. American Libraries, 3(9): 987-99; October, 1972.
(Williams & Wilkins)

Smith, Roger H. A copyright commission in the public interest. Publishers' Weekly, 199(20): 35; May 17, 1971.
(news)

Sophar, Gerald J. Williams & Wilkins vs. the United States Government. Newsletter of the American Society for Information Science, 9(5): 1,4; September-October, 1970.
(Williams & Wilkins)(news)

Sordelli, Luigi. Diritto di autore, riproduzione di opere mediante la C.D. reprografia ed interessi personali e collettivi alla cultura. Il Diritto di Autore, 46(4): 496-505; October-December, 1975.
(in Italian)
(international, Italy)

An analysis of the 1941 Italian copyright law is presented in the perspective of the articles pertinent to the scope of permissible photocopying. Articles 65-70 deal with photocopying for personal use and in the public interest. These provisions are also discussed with respect to the continuing UNESCO/WIPO sponsored meetings on the subject, which have been underway since 1973.

Spreitzer, Francis F. Developments in copying, micrographics, and graphic communications, 1971. Library Resources and Technical Services, 16(2): 135-54; Spring, 1972.

Three of the four significant developments of the year focus on micrographics; the fourth is copyright and full-size copying in the library milieu. At this point, "full-size photocopying in the library is quantitatively and functionally much more important than micrographics." The first part of this article deals with new copying machines, library copying practices and needs, and the controversy surrounding copyright as reflected in concern in the publishing industry and proposals for copyright legislation.

Spreitzer, Francis F. Developments in copying, micrographics, and graphic communications, 1972. Library Resources and Technical Services, 17(2): 144-67; Spring, 1973.

An extensive review article appears in six sections: 1) general review; 2) full-size copying; 3) microforms; 4) microform equipment; 5) audiovisual and facsimile transmissions; and 6) publications, research and professional activities. In the general review section, approximately four pages are devoted to an assessment of the Williams & Wilkins case and its impact upon photocopying in libraries. The slant is unfavorable toward publishers in general and Williams & Wilkins in particular, especially in regard to the economics of publishing. "The crux of the matter appears to be some publishers' discovery that library photocopying is not so much a threat to be feared as a moneymaking opportunity to be exploited."

Spreitzer, Francis F. Developments in copying, micrographics, and graphic communications, 1973. <u>Library Resources and Technical Services</u>, 18(2): 151-70; Spring, 1974.

In the first part of this review of technological developments in reprography, the author discusses the ascendancy of legal over technological developments as the prime area of interest during 1973. The complex issues centering on the introduction of S. 1361 for copyright revision, especially Section 108 on fair use, are detailed with respect to libraries, library associations, and educational institutions; the distinction is drawn between copying of print versus non-print materials. The Williams & Wilkins case is reviewed and comment made on the Court of Claims decision. The remainder of the article is devoted to the review of materials and equipment in micrographics and other technological developments.

Statement of the Association for Educational Communications & Technology. Howard B. Hitchens, Executive Director, to the Subcommittee on Courts, Civil Liberties and the Administration of Justice to the House Judiciary Committee on H.R. 2223. May 15, 1975. 7p.

Steinmetz, Helmut. Gedanken zur photomechanischen Vervielfältigung. <u>Österreichische Blätter für Gewerblichen Rechtsschutz und Urheberrecht</u>, 21(3): 53-56; May-June, 1972. (in German)
(international, Austria)

The problems of photomechanical reproduction of copyrighted works are discussed with respect to Austrian copyright law. The author attempts to clarify section 15, which covers the author's exclusive rights to reproduction of his works, and section 42, which limits the author's rights as above in connection with reproductions for personal use.

Stevenson, Iris C. Doctrine of fair use as it affects libraries. Law Library Journal, 68(3): 254-73; August, 1975. (Williams & Wilkins)

Stinespring, E. Book and journal permissions. Scholarly Publishing, 6: 69-74; October, 1974.

Stuart-Stubbs, B. Purchasing and copying practices at Canadian university libraries: two studies performed for the Canadian Association of College and University Libraries, Committee on Copyright Legislation.. Ottawa: Canadian Library Association, 1971. 46p. (international, Canada)

This is a report on a study conducted by obtaining information from member libraries of the Canadian Association of University Libraries between January and April, 1971 in order to determine: 1)pattern of library spending, indicating the development of their collections; and 2) the effects of direct purchasing of materials from country of origin available through Canadian agents. From the part of the study devoted to duplication practices covering numbers of photocopying machines, numbers of copies, type of material copied, and for whom copies were made, it was hoped to gain information on areas impacted by copyright and which would be affected by copyright law revision.

Studiecommissie Fotomechanische Reproductie. About photomechanical reproduction. Report of the Mixed Study Committee for the study of this question. Inter-auteurs, no. 179: 121-37; 2d quarter, 1970. (international, Netherlands)

The report of the mixed study committee to study the problem of reprography and copyright in the Netherlands is presented. They state that there has been little in the literature regarding solutions to the problem except in the German Federal Republic where a general agreement exists under the German copyright act of 1965, and a suggestion by the Committee to Investigate Copyright Problems (CICP) in the United States for a clearinghouse. Both of these aspects are discussed at length relating to their applicability to Netherlands law.

Subcommittee of the Executive Committee of the International Union for the Protection of Literary and Artistic Works on Reprographic Reproduction sitting together with the Subcommittee of the Intergovernmental Copyright Committee on Reprographic Reproduction (Washington, D.C., June 16-21, 1975). Copyright, 11(8): 159-76; August, 1975. (international)

The agenda of the joint meeting is reported on the following topics: 1) remuneration for and control of reprographic reproduction; 2) statutory licensing proposals; 3) surcharges on reprographic equipment; 4) private uses of copied material; 5) uses by non-profit institutions (libraries, archives, documentation centers, public scientific research institutions); 6) uses by educational institutions; 7) commercial organizations; 8) special problems of developing nations. Stating that the reprographic problem does not occur uniformly within each nation, the committee declined to offer a uniform solution on an international scale, and proposed that each member country adopt whatever measures deemed appropriate, and that where large-scale copying occurs, the countries should encourage establishment of collection agencies for remunerative purposes.

Subcommittee of the Intergovernmental Copyright Committee on Reprographic Reproduction of Works Protected by Copyright (sitting together with the Subcommittee of the Executive Committee of the Berne Union). Washington, D.C., 16-21 June 1975. Report. UNESCO Copyright Bulletin, 9(2/3): 18-48; 1975. (international)

For a review of this report see item immediately above.

Sword, Larry F. Photocopying and copyright law - Williams & Wilkins Co. v. United States (487 F 2d 1345): how unfair can "fair use" be? Kentucky Law Journal, 63(1): 256-78; 1974-1975. (Williams & Wilkins)

-T-

Tallman, Johanna E. An affirmative statement on copyright debate. Journal of the American Society for Information Science, 25: 145-50; May-June, 1974.

Information on government supported research and publication should be disseminated at a minimal cost and maximal accessibility "at no extra (author's emphasis) profit to the publisher." Only if subscription costs keep rising will circulation decline; Congress should continue protection against infringement.

The teacher and the copyright law. NJEA Review, 43(5): 36-37; January, 1970. (news)

Technology, copyright, government and the Court. Information Retrieval & Library Automation Newsletter, 7: 7-9; April, 1972. (Williams & Wilkins)(news)

Thorpe, James. The use of manuscripts in literary research. New York: Modern Language Association of America, May, 1974. 40p.

This is a pamphlet prepared as a guide to the use of manuscripts for the scholar in literary research. Of interest here are the sections on access to the documents dealing in part with photocopying and permission to publish, and literary property rights, covering copyright and common law copyright (protection of unpublished material), infringement, literary rights in other countries, and under international copyright.

Tyerman, Barry W. The economic rationale for copyright protection for published books: a reply to Professor Breyer. UCLA Law Review, 18: 1100-25; June, 1971.
(This article subsequently appeared in Bulletin of the Copyright Society, 19(2): 99-128; December, 1971 and Copyright Law Symposium (ASCAP), 21: 1-34; 1974)

The author attacks Professor Breyer's thesis (see Breyer, 1970),that the case for maintaining copyright protection in books is weak,with a point by point rebuttal. He first deals with the concept of the conditions of the publishing industry without copyright in regard to costs to publisher and benefits to society, then turns to the economic points in detail relative to maintaining book production, fear by reproducers of publisher retailiation, potentially profitable publisher-copier coexistence, and the risk to authors without copyright protection. He also questions whether society would truly be benefited by doing away with copyright, and concludes that the "abolition of copyright protection for published books would have a serious detrimental effect on authors, publishers, and our society...".

Ulmer, Eugen. Automatic and in particular, computerized information and documentation systems and the copyright law. Copyright, 11(12): 239-46; December, 1975. (international)
(Original German language article appears under Kolle, Gert and Ulmer, Eugen in Gewerblicher Rechtsschutz und Urheberrecht, no. 3: 108-16; March, 1976)

This is a discussion of copyright problems arising from computer storage and retrieval of copyrighted works,first presenting background information on the nature of material stored in these systems, and the national and international information and documentation systems that have begun to appear in response to the rapid increase in information, and attempts to provide mechanisms for furthering access to this information. Discussed at some length are the uses of microforms in these ISAR systems. Several types of documentation centers are presented, both on a central scale and on a private scale. The author then relates these developments to the copyright laws detailing the basis for protection, the limits of this protection with reference to reprography. He concludes that considering all aspects, at the minimum some sort of remuneration to the owners is required, and that in considering attempts to regulate information systems, two problems must be attacked: 1) facilitation of access to material linked to remuneration for such access; and 2) "a judicious harmonization" between publishing output and supply of information by technologically created systems.

United Nations Educational, Scientific and Cultural Organization. General Conference. 17th session, Paris, October-November, 1972. UNESCO Copyright Bulletin, 7(2/3): 24-44; 1973. (international)

Resolution concerning photoreproduction. Report on the advisability of adopting an international regulation concerning the photographic reproduction of copyrighted works.

The general conference resolution recommends the preparation of an international instrument for the purpose of setting guidelines in the study of the photocopying problem and judges that this recommendation be prepared and sent to each of the member states. This resolution is followed by a report stemming from the preceding year's meeting on the desirability of an internation regulation on photocopying of copyrighted works. The report encompasses background studies, extent and scope of copyright protection, reproduction for private use, copying by libraries, provision of copies to third parties, copies for educational purposes, and the methods by which the regulation of photographic reproduction should be attempted.

University of Maryland. College of Library and Information Services. Proposed solutions to the problem of photocopying copyrighted material: a survey. College Park, Md.: April, 1975. 35p. (unpublished)

Report by the graduate class in Operations Research for Library and Information Science: LBSC 858H. Faculty adviser, Laurence B. Heilprin.

Fourteen possible solutions to the copyright/photocopying problem were analyzed and tested against a set of fifteen criteria. The evaluations were made in light of five "parties of interest" whose opinions must be considered in the controversy: authors, publishers, technological.innovators (manufacturers of reprographic equipment), intermediate disseminators (librarians, educators), and users themselves. The solutions with the highest scores included: contractual agreements between publishers and users; two-tier subscription prices for periodicals; statutory option offering two classes of copyright (in Appendix A), and a non-profit voluntary clearinghouse.

University of Maryland. College of Library and Information Services. Copyright clearinghouse: in-depth study of a proposed solution to the copyright photocopying problem. College Park, Md.: Spring, 1976. 39p. (unpublished)

Report by the class in Operations Research for Library and Information Science: LBSC 858H. Faculty adviser, Laurence B. Heilprin.

This paper is a follow-up to the preceding year's entry (see University of Maryland above) of proposed solutions to the copyright problem, presenting an in-depth study of the solution that met the most requirements of the earlier study - a copyright clearinghouse. The clearinghouse is envisioned as a non-profit, voluntary association (as opposed to statutory) between copyright owners and users, with each group sharing equally in its direction. Copying privileges are granted to the clearinghouse in the form of individual contracts between the publishers and the clearinghouse. Additional contracts between the clearinghouse and user institutions provide for assessment of fees for copying based on results of periodic sampling studies. The sampling insures fair distribution of costs based on rate of use, and reduces the burdens of record keeping by the user to a minimum. The overall administrative organization of the clearinghouse, details of sampling methodology, a mathematical model of clearinghouse economics and rate setting, and the requirements necessary for implementation of the model, are discussed in detail.

-V-

Vander Ploeg, Jon. Copyright law and library photocopying: striking a balance between profit incentive and the free dissemination of research information. Indiana Law Journal, 48(3): 503-15; Spring, 1973. (Williams & Wilkins)

Van Tongeren, E. The effect of a large-scale photocopying service on journal sales. Journal of Documentation, 32(3): 198-204; September, 1976.

This is a response from a publisher's viewpoint to the results of a survey of photocopying practices at the British Library Lending Division (see Line & Wood, 1975). The author refutes the claims of the earlier article that state little, if any, impact of photocopying on sales. He discusses nine specific points from the survey and concludes that, contrary to the earlier claims, photocopying presents "a serious and immediate threat" to scientific and technical publishing. Laws must be passed to control photocopying, meaning that "entitlement to make single copies as a matter of right (author's emphasis) should be removed from the existing law."

Viggiano, Frank A., Jr. A study of various educational media personnel and their practices regarding the reproduction of copyrighted works. Stout, Menomonie: University of Wisconsin Graduate College, August, 1974. 67p.

This is a report based on information gained from an "opinionnaire" distributed to attendees at the ninth Annual Educational Media and Technology Conference, July 22-24, 1974 at the University of Wisconsin. The questionnaire was designed to uncover attitudes on reproducing copyrighted material in seven categories covering twenty-four hypothetical work situations. The responses were broken down by category and by professional status of the person answering. In a majority of the twenty-four situations, the results indicated that copying would have occurred. Information is given on professionals most/least likely to copy all types of copyrighted material, and copyrighted materials most/least likely to be copied. Also included in the report is a brief review of the current copyright law and the status of legislative attempts toward revision.

-W-

Wagner, Susan. Government loses photocopying suit. Publishers' Weekly, 201 (9): 48-49; February 28, 1972. (Williams & Wilkins)(news)

Wagner, Susan. Midnight oil burns after Williams & Wilkins rule. Publishers' Weekly, 202(7): 29-30; August 14, 1972. (Williams & Wilkins)(news)

Wagner, Susan. Photocopying talks collapse; new solutions sought. Publishers' Weekly, 202(24): 20; December 11, 1972. (Williams & Wilkins)(news)

Wagner, Susan. A visist with Williams & Wilkins. Publishers' Weekly, 203(14): 22-23; April 2, 1973. (Williams & Wilkins)(news)

Wagner, Susan. Copyright: educators dig in their heels. Publishers' Weekly, 204(7): 32-33; August 13, 1973. (news)

Wagner, Susan. Looking ahead at the Library of Congress. Publishers' Weekly, 205(24): 43; June 17, 1974. (news)

Wagner, Susan. "Fair use" carefully defined in copyright revision bill. Publishers' Weekly, 205(25): 31-32; June 24, 1974. (Williams & Wilkins)(news)

Wagner, Susan. Library community modifies its position on "fair use". Publishers' Weekly, 206(2): 20; July 8, 1974. (news)

Wagner, Susan. Are the Soviets engaging in unauthorized photocopying? Publishers' Weekly, 206(8): 22; August 19, 1974. (news)

Wagner, Susan. Copyright vs. fair use: no agreement in sight. Publishers' Weekly, 207(22): 25-26; June 2, 1975. (news)

Wagner, Susan. NCLIS study will explore photocopying. Publishers' Weekly, 208(17): 18-19; October 27, 1975. (news)

Wagner, Susan. "Fair use" section passes in House mark-up session. Publishers' Weekly, 209(11): 18; March 15, 1976. (news)

Walsh, John. Journals: photocopying is not the only problem. Science, 183: 1274-77; March 29, 1974. (news)

Warasilia, R. Case of morality with legal overtones. Physics Teacher, 13: 198; April, 1975. (news)

Weil, Ben H. Copyright developments. In: Annual Review of Information Science and Technology, ed. by Carlos J. Cuadra. Washington, D.C.: American Society for Information Science, 1975. pp. 359-81.

The paper is presented as a follow-up to Beard, 1974, and reviews the current developments in the copyright issue. Discussion is broken down by: 1) judicial developments mentioning the Williams & Wilkins case; 2) legislative developments reviewing Congressional efforts in passing the copyright revision bill; 3) information community developments focusing on the National Commission on Library and Information Science and the Copyright Office, the private sector, planning, networks, publishing and abstracts; and 4) international and foreign developments.

Weinberg, Louise. The photocopying revolution and the copyright crisis. Public Interest, 38: 99-118; Winter, 1975.

The tremendous advances in reprographic technology leading to increased use of various photocopying services are discussed from the point of view of their impact on and potential threat to the publishing industry. The author reviews the current copyright law and recounts its inadequacies in this regard along with suggested remedial legislation in terms of proposals for licensing.

Weinstock, R. Case of morality with legal overtones: a reply. Physics Teacher, 13: 325; September, 1975. (news)

Response to Warasilia, 1975

White, Anthony G. Copyrights: a selected bibliography (Exchange bibliography, no. 686). Monticello: Council of Planning Librarians, 1974. 9p.

This is a bibliography perhaps more suitable for those interested in a general background overview of literature in all areas of copyright. It is a simple alphabetically listed set of approximately one hundred citations from the early 1900's forward with only two or three from the 1970's, and no annotations. The slant of the collection is toward the impact of technology on copyright.

Wigren, Harold E. The teacher and the copyright law: a response. Journal of Reading, 17(1): 16-19; October, 1973. (news)

Response to Caughran, 1973.

Wildberger, M.E. When copyright comes to curriculum. Media and Methods, 12: 70-71; October, 1975. (news)

Williams & Wilkins: libraries win round one. American Libraries, 5(2): 60-61; February, 1974. (Williams & Wilkins)(news)

The Williams & Wilkins Company v. the United States (decided November 27, 1973). Washington, D.C.: Court of Claims, no. 73-68; November, 1973. 104p. (Williams & Wilkins)

Williams & Wilkins Company v. the United States: report of the Commissioner to the Court. Washington, D.C.: Court of Claims; February 16, 1972. 63p. (Williams & Wilkins)

Wilson, Samual A., Jr. The United States Court of Claims on copyright: Williams & Wilkins Company v. the United States (487 F 2d 1345) in perspective. Mercer Law Review, 26(4): 1401-15; Summer, 1975. (Williams & Wilkins)

Winkler, Karen J. Copyright. College Store Journal, 42(6): 98-101; October/November, 1975.

The author discusses the issues involved in the revision of copyright bills approved by the U.S. Senate. Of particular interest are the continuing conflicts between publishers and educators and librarians regarding photocopying of copyrighted material, especially articles in periodicals for use in classrooms or for interlibrary loan.

Winn, Karyl. Common law copyright and the archivist. <u>American Archivist</u>, 37 (3): 375-86; July, 1974.

Dealing with the difficulties of interpretation of common law copyright (the rights of authors to exclusive use and control of literary creations before publication) regarding copying of donated materials for limited use, the relationship to fair use and the provisions of S. 1361, which would protect unpublished writings more specifically under statute than is now the case under common law is explored. The author discusses the provisions allowed under S. 1361 for reproductions for private study, scholarship or research and under limited conditions, the making of facsimiles for preservation and deposit in repositories.

Working Group on Reprographic Reproduction of Works Protected by Copyright. Paris, May 2-4, 1973. Report. <u>Copyright</u>, 9(9): 172-80; September, 1973. (Annexes 179-80). (international)

This is a report summarizing the meeting of the group with the frame of reference to "prepare an international instrument on the question of photographic reproduction of works protected by copyright" in the form of recommendations. The recommendations are contained in the Annex A of the report and state that: 1) authors are entitled to fair remuneration for reprographic reproduction of their copyrighted works; 2) individuals may make single copies of single articles from periodicals or reasonable portions of other works; 3) such copies as allowed in 2) may be made for an individual by a library or documentation center; 4) blanket licenses should be negotiated between educational authorities and an organization representing authors and publishers for purposes of making limited numbers of copies of copyrighted works in teaching; 5) licensing systems as in 4) also should be made available to public administrators and commercial enterprises; 6) developing countries are not obligated under copyright conventions to provide remuneration to the same extent as outlined in the preceding paragraphs; 7) reprography includes any means by which facsimile reproductions are made.

Wright, Christopher. A "third" force in photocopying dispute? Library Resources and Technical Services, 18(4): 341-42; Fall, 1974. (news)

-Y-

Yanow, Lilli A. Copying, the reconciliation of conflicting interests: an analysis of Williams & Wilkins Co. v. United States (172 USPQ 670). Rutgers Journal of Computers and the Law, 3(2): 328-44; 1974. (Williams & Wilkins)

-Z-

Zanden, J.W. van der. Mag de Octooiraad een voor eigen gebruik gemaakte fotocopie van een tijdschriftartikel aan een gemachtigde ter inzage geven? Bijblad bij de Industriele Eigendom, 41(1): 11-15; January, 1975. (international, Netherlands) (in Dutch)

The question is posed as to whether the Patent Council in the Netherlands may make copies of periodical articles for personal use. The articles are used in studying current patent applications and court litigation. Comparison is made with the amendments in the July, 1974 Dutch copyright law and the current law with respect to photocopying.

SUBLIST OF

SELECTED, ANNOTATED

BIBLIOGRAPHY, ON

WILLIAMS AND WILKINS vs. THE UNITED STATES

by

Peri L. Schuyler

American Association of Law Libraries. 65th Annual Meeting, July 2-6, 1972. General session on copyright. Law Library Journal, 65(4): 443-53; November, 1972.

The general session on copyright was given over to discussion of the Williams & Wilkins case. A brief review of the case and the Commissioner's findings for the plaintiff, Williams & Wilkins, was presented by Carelton W. Kenyon, who reluctantly sided with the Commissioner, although indicating that there were mitigating facts making the decision a bad one. He indicated that the only solution in his opinion is through legislation and court decisions. A strong argument against the decision was made by Julius J. Marke, who stressed the negative ramifications on the extension of library services, the possibilities of libraries being legal defendants in many more case, the increased costs of journal subscriptions to institutions, and the general financial squeeze encountered by all libraries. Further discussion by other panel members was offered on possible solutions and roles of libraries and library associations in reaching these solutions.

Benjamin, Curtis G. A hard look at the new Williams & Wilkins decision; the Court of Claims ruling presents a somber prospect for a wide variety of specialized publishers. Publishers' Weekly, 205(10): 33-34; March 11, 1974. (news)

Benson, Stuart W., III. Technological piracy in the reprographic revolution. University of Pittsburgh Law Review, 36: 153-73; Fall, 1974.

The author outlines the United States Copyright Law development and the evolution of the judicial "fair use" doctrine and then proceeds to a discussion of the Williams & Wilkins case with respect to the concepts of fair use and infringement. He concludes that the Court of Claims decision in favor of NIH and NLM was "unwarranted judicial 'legislation'", claiming that in the infringement issue, the photocopying performed by the libraries was an "abridgement of the copyright owner's exclusive right to copy", and that this copying did not constitute "fair use". He discusses several solutions including the reserving of photocopying rights via government grants, various royalty collection systems and judicial solutions.

Berkowitz, Linda L. Copyright law - fair use - photocopies: Williams & Wilkins Co. v. the United States. New York Law Forum, 20(4): 819-31; Spring, 1975.

The author presents a critical analysis of the results of the Supreme Court decision which upheld the Court of Claims ruling that defendants, NIH library and NLM, were not guilty of infringing plaintiff Williams & Wilkins. The Supreme Court actually came to no decision, but arrived at a tie vote with one abstention resulting in upholding by default. The author concludes that this will begin a permissive trend in the use of copyrighted works, which will work to unfair burdens on copyright owners, especially by increasing expenses due to loss of deserved revenues.

Brewer, Karen. A healthy approach to copyright problems. Publishers' Weekly, 202(11): 21-22; September 11, 1972. (news)

Refers to Doebler, Publishers' Weekly, 202(4): 52-54; July 24, 1972.

Bungeroth, Marc R. Copyright - photocopying as fair use. Boston College Industrial and Commercial Law Review, 16(1): 141-50; November, 1974.

Carnahan, William H. Fair use and the libraries. Air Force Law Review, 17(3): 41-51; Fall, 1975.

The article summarizes both the Commissioner's report in the Williams & Wilkins case and the findings of the Court of Claims in a step-by-step analysis of the rulings. He concludes that the real issue is "copying for private use, not general publication" and presents as "fundamental fact ... that the Copyright Act has nothing to do with private uses of copyrighted material by scholars..". The Court pointed out that the libraries had made no attempt at profit, had established strict limitations on copying, that their practices closely resembled Library of Congress practices, that libraries had been copying for at least sixty-five years with no complaints until approximately fifteen years ago. The author feels that of far greater concern to libraries is the impact of the current revision of the Copyright Law underway in Congress, particularly Section 108 dealing specifically with library photocopying rights.

Copyright--fair use--photocopying and distribution of copyrighted journal articles by government library constitutes actionable infringement. Vanderbilt Law Review, 25(5): 1093-107; October, 1972.

Copyrights--infringement by photocopying. Texas Law Review, 51(1): 137-48; December, 1972.

Court ruling threatens photocopying. Chemical & Engineering News, p. 14; March 6, 1972. (news)

Culliton, Barbara J. Photocopying habit in jeopardy. Science, 176: 265; April 21, 1972. (news)

Dembart, Lee. Photocopying splits publishers, libraries. New York Times, at p. 30, columns 1-4 (city edition); October 29, 1974. (news)

Diamond, Sidney A. Split Supreme Court decision leaves photocopying issue clouded. (What's doing on the legal front?) Advertising Age, 46(19): 50; May 12, 1975. (news)

Doebler, Paul. IIA discusses the copyright dilemma. Publishers' Weekly, 202(4): 52-54; July 24, 1972. (news)

Fotokopieren in den USA -- Schutzlosigkeit von Autoren und Verlegern? Gewerblicher Rechtsschutz und Urheberrecht, Int. Teil, no. 6: 246-51; June, 1974. (in German)

A German language article covers the Court of Claims decision in the Williams & Wilkins case. The article is in two parts: first an analysis of the Court's decision followed by a critical commentary on its potential impact, concluding with the hope that when the case reaches the Supreme Court, that Court will ultimately solve the problem of library photocopying.

Goldstein, Paul. Private consumption of public goods: a comment on Williams & Wilkins Co. v. U.S. (487 F 2d 1345). Bulletin of the Copyright Society, 21(3): 204-13; February, 1974.

This article is a critical comment on the Court of Claims decision ruling for the defendants, NIH and NLM in the Williams & Wilkins case. The author divides his treatise into two parts - the first a survey of the potentially negative economic impacts of the decision, and the second the possible consequences of the Court's avowal of the doctrine of judicial restraint. The two main economic aspects are, in the author's opinion, in the realm of financial harm to the copyright owner and in the effect of imposing transaction costs to the user. He concludes that the Court ignored the distinction between "general rule and practical exception and, specifically, mistakes a past consequence of market expedience for the pronouncement of a legal exception to the (Copyright) Act's policy of general distribution (of publishers' costs)".

Grannis, Chandler B. Copyright protects the public interests. Publishers' Weekly, 202(21): 48; November 20, 1972. (news)

Hart, Bruce. Copyrights: concurrence, revision, and photocopying - Williams & Wilkins Company v. the United States (487 F 2d 1345): a "holding operation" by the United States Court of Claims. Dickinson Law Review, 79(2): 260-308; Winter, 1975.

The author presents a three-part commentary on the history and status of copyright law revision, with emphasis on the Williams & Wilkins case in regard to the photocopying quandary. Part 1 deals with development and analysis of concurrence and revision policies; part 2 discusses the current revision of the bill now underway in Congress, and outlines earlier attempts that met with obstruction; and part 3 analyzes both the Commissioner's report to the Court of Claims holding for plaintiff, Williams & Wilkins, and the Court of Claims ruling overturning the Commissioner and finding for the defendants, NLM and NIH. It considers possible impacts of the case on the warring parties in the photocopying dispute.

Harvell, Michael. Copyright--fair use--government photocopying does not infringe journal copyright. Boston University Law Review, 54: 689-701; May, 1974.

Objection is made to the Court of Claims decision in reversing the holdings of the Commissioner and finding NIH and NLM not guilty of infringement. The defense of fair use was improperly used in the author's opinion, not only because by relying on this clause, no decision could be made as to whether the photocopying would be an infringement in the absence of the clause or whether it could ever be construed as infringement, but because it overemphasized the importance of"potential injury" to medical research and placed too heavy a burden on the plaintiff, Williams & Wilkins, to prove actual damages, rather than potential harm. It is felt by the author that this approach departs from traditional "protection of (copyright) holder's financial interests" and focuses instead on balancing potential harm to both parties at issue.

Hollingsworth, Robert J. Copyright--libraries--unauthorized library photocopying of articles from copyrighted medical journals constitutes copyright infringement. University of Cincinnati Law Review, 41(2): 511-19; 1972.

The author presents an in-depth analysis of the conflict between "fair use" and societal demands for free flow of information and the protection of the author and publisher guaranteeing exclusive rights of reproduction. The Commissioner held that the defense of single-copy photocopying was "illusory and unrealistic", observing that multiple requests in fact generated multiple copies. Although the article deals somewhat sympathetically with the defendants, NIH and NLM, in the case for library photocopying, it concludes that the decision for plaintiff, Williams & Wilkins, is reasonable at this time and feels that the case will generate solutions to the conflict between publisher and public.

Justices to decide if libraries must pay copyright for photocopying items. Wall Street Journal, 183(103): 7, columns 3-4; May 29, 1974. (news)

Keenan, Stella & Stillman, Mary E., issue ed. Information Retrieval Cooloquium, 9th. Philadelphia, Pa., May, 1972. Panel on copyright. The copyright controversy: issues and opinions. Drexel Library Quarterly, 8(4): 377-602; October, 1972.

The entire issue is divided into panel presentations, group discussion, and attached relevant appendices. Members of the panel who first spoke individually and then participated in the group discussions, were: Irwin Karp speaking for the author; Julius Marke speaking for the librarian: and Benjamin Weil speaking as an "activist". Ms. Keenan presented her views in a general introduction to the panel. The attached appendices (which begin on p. 415) are: Appendix A - Brief Amicus Curiae of the American Association of Research Libraries, Medical Library Association, and American Association of Law Libraries in the United States Court of Claims; Appendix B - Brief Amicus Curiae of the Authors League of America, Inc. in the United States Court of Claims; and Appendix C - Report of the Commissioner to the Court of Claims.

Legality of photocopying copyrighted publications. Williams & Wilkins Co. vs. United States. William & Mary Law Review, 13(4): 940-48; Summer, 1972.

The author presents a brief outline of the copyright law in terms of its securing protection for rights of authors, the scope of infringement, and a particular emphasis on the difficult area of "fair use", with examples. The article then turns to the Williams & Wilkins case, the only example at the time of the article of a court case involving reprography and its relation to fair use. The impact of "the reprographic revolution" on publishing is discussed with mention of the various solutions currently suggested in the literature and concluding with the hope that future court decisions will uphold the court's decision in finding for the publisher.

MacKenzie, John P. Supreme Court faces issue of copyright vs. copying. The Washington Post, p. A2, columns 1-3; December 18, 1974. (news)

McCormic, Marilyn G., comp. The Williams & Wilkins Case: The Williams & Wilkins Co. v. the United States. Volume I. New York: Science Associates/International, 1974. 275p.

This volume is a collection of all the documents filed with the Court of Claims and documents filed with the United States Supreme Court in the case of

Williams & Wilkins vs. the United States (NIH and NLM). Included are the report of the Commissioner to the Court of Claims, amici curiae briefs for plaintiff and defendant, plaintiff and defendant briefs (contains briefs presented both to the Court of Claims and to the Supreme Court), and the decision reversing the Commissioner by the Court of Claims. Also in the volume are an introduction by Nicholas Henry and a statement by Williams J. Passano, then chairman of the board of Williams & Wilkins. Volume II will present the decision of the Supreme Court (upholding the Court of Claims by a tie vote) and its related papers.

Murdock, Kent. Williams & Wilkins Co. v. U.S. (487 F 2d 1345): library photocopying of copyrighted materials. Utah Law Review, 1974: 127-44; Spring, 1974.

Although an objective facet-by-facet analysis of the Court of Claims ruling which held for NLM and NIH overturning the Commissioner's decision, the author feels that the issue is by no means settled, and that if the Court had upheld the Commissioner by ruling infringement, a private accommodation of competing interests would have been forced. This accommodation in his opinion might have been "the most viable manner of protecting the public interest." The present ruling is felt to sacrifice copyright owners' rights which were not reflected in the Court's opinion.

Murphy, James P. The legal effects of Williams & Wilkins. Publishers' Weekly, 202(24): 11-12; December 11, 1972. (news)

Refutes Passano in article Publishers' Weekly, 202(20): 17-18; November 13, 1972.

Nasri, William Z. Impact of reprography on the commercial publication of scientific and technical periodicals with particular reference to the case of "The Williams & Wilkins Co. vs. the United States". Ph.D. dissertation. Pittsburgh: University of Pittsburgh, 1975.

The author presents a detailed study centering around the claims of publishers of scientific, technical, and especially medical journals as to potential subscription loss due to photocopying. An in-depth analysis of the Williams & Wilkins case is followed by the methodology the author used in attempting to determine the impact of reprography upon publishing. It had been originally hoped that the publishers themlseves would be able to provide such data, but precise data were unavailable. The effort then centered on users of four journals published by Williams & Wilkins, which were, as it turned out, also the same four chosen by that publisher to plead their original suit. Questionnaires were sent to a random sample of users identified mostly from medical association membership lists to determine reasons for dropping or failing to subscribe to any of the four journals, and if photocopying was the most important factor in this regard. The results indicated a negligible effect and that library service served as a positive reinforcement of an already made decision. Primary reasons given for subscription loss were either time constraints, financial considerations, or a change in subject interest.

North, William. Williams & Wilkins. The great leap backward. American Libraries, 3(5): 528-30; May, 1972. (news)

Nyswonger, J. Stephen & Kennedy, Craig. Copyright: library use of photocopiers. Washburn Law Journal, 12(2): 235-40; Winter, 1973.

Oettinger, Elmer R. Copyright and the librarian. North Carolina Libraries, 31 (4): Conference issue, 1973.

The author reviews the problems of photocopying in libraries centered around the Williams & Wilkins case. Exploring the decisions of both the Commissioner, who ruled for Williams & Wilkins, and the Court of Claims, who ruled for the defendants, NIH and NLM, he concludes that at best a compromise between the warring factions - publishers on one side and educators and librarians on the other - must arise to accommodate some portion of the needs of each.

O'Neill, Judith D. Copyright - extensive photoduplication of copyrighted scientific journals does not constitute copyright infringement. University of Baltimore Law Review, 3(2): 306-17; Spring, 1974.

The author refutes the Court of Claims reversal of the Commissioner's decision finding for the defendants, NLM and NIH. The Court's opinion based on the defense of "fair use" was "clearly erroneous as the doctrine of fair use cannot be examined where whoesale taking exists." The author states that each article, rather than being considered as part of a whole journal, is an individual work in itself; thus copying of a single article represents wholesale copying of the entire work. Issue is also taken with the Court's claim that Congress bears responsibility for legislation rather than the courts, on the basis that it is the duty of the Court to interpret law, and that the legislature has been "unable to cope with the problem statutorily since its inception."

Passano, William J. How photocopying pollutes sci-tech publication. Publishers' Weekly, 197(5): 63-64; February 2, 1970. (news)

Passano, William J. A statement from Williams & Wilkins to the editors, Publishers' Weekly, by William J. Passano, chairman of the board, Williams & Wilkins Co. Publishers' Weekly, 202(20): 17-18; November 13, 1972. (news)

Passano, William J. A publisher's view of photocopying. American Libraries, 5(5): 222; May, 1974. (news)

Photocopying for researchers held legal. Chemical & Engineering News, 51(50): 12-13; December, 1973. (news)

Photocopying: high court tie vote leaves issue to Congress. Science, 187: 934; March 14, 1975. (news)

Reitz, Norman E. Williams & Wilkins: (Williams & Wilkins Co. v. U.S. 172 USPQ 670): the impact of technology on copyright. Los Angeles Bar Bulletin, 48(12): 445-50, 468-72; October, 1973.

An analysis is made of the long-range implications of the report of the Commissioner to the United States Court of Claims in the Williams & Wilkins case, in which the Commissioner found NLM and NIH library guilty of infringement of copyrighted material owned by Williams & Wilkins. The author discusses the current reprographic practices as they affect publishers and copyright owners and the impact of the present ruling on libraries and research organizations.

Rosenfield, Harry M. Customary use as "fair use" in copyright law. Buffalo Law Review, 25(1): 119-40; Fall, 1975.

This is an article adapted from the Brief Amicus Curiae submitted by the author on behalf of the National Education Association in the Williams & Wilkins case. He believes that the current copyright revision bills under study in Congress will result in the reversal of common law doctrine on fair use which grants customary non-profit use of copyrighted materials for non-commercial education, research, and scholarship is "fair use". The author suggests appropriate modifications in the attempted revisions will be necessary in order to forestall this negative possibility.

Shaw, Ralph R. Williams & Wilkins v. the U.S.; a review of the Commissioner's Report. American Libraries, 3(9): 987-99; October, 1972.

The author presents a critical review of the Commissioner's report to the Court of Claims in which NIH and NLM were found guilty of infringement in the Williams & Wilkins case. He states that although the original issue was simply a copyright infringement suit, far broader implications have arisen concerning scholars' rights to make notes or copies regardless of the means for their own personal use, the rights of libraries to act as agents for scholars in making single copies at their requests, the state of medical publishing, and dangers of government control. The author declares that before a decision can be reached as to infringement of copyright, two fundamental points must be settled: 1) is there a valid copyright of the journal as a whole, and does plaintiff own that copyright? and 2) if plaintiff owns a valid copyright of the whole journal and its issues, have the individual articles been assigned to him? The bulk of the article is given over to rebuttal of the Commissioner's findings with respect to these points, and to the issue of public versus private use.

Sophar, Gerald J. Williams & Wilkins vs. the U nited States Government. Newsletter of the American Society for Information Science, 9(5): 1,4; September-October, 1970. (news)

Stevenson, Iris C. Doctrine of fair use as it affects libraries. Law Library Journal, 68(3): 254-73; August, 1975.

An extensive and thoughtful analysis of the fair use clause is set forth with especial reference to the case of Williams & Wilkins. This was the first case of such nature brought before courts, and the first suit for copyright infringement against the government since it waived its immunity to such suits. The distinction, emphasized by the Court of Claims between NLM's and NIH's reproduction of copyrighted works, and those in other prior cases, was that the libraries produced copies for personal use rather than for incorporation into some other work for public consumption. Extensive review of various legal rulings on the issue of fair use is also presented with a conclusion that discusses various solutions and their implications for public and private interests.

Sword, Larry F. Photocopying and copyright law - Williams & Wilkins Co. v. United States (487 F 2d 1345): how unfair can "fair use" be? Kentucky Law Journal, 63(1): 256-78; 1974-1975.

This paper is an analysis of the Court of Claims decision which reversed the findings of the trial commissioner in the Williams & Wilkins case, and found for the defendants,. NLM and NIH. Highly critical of the Court of Claims decision, the author concludes that the photocopying practices of the respective libraries did indeed produce harm to the publisher which might therefore cause significant loss in potential income, and that publishers of medical journals must have protection under the law in order to maintain the economic viability necessary to continue publishing such journals.

Technology, copyright, government and the Court. Information Retrieval & Library Automation Newsletter, 7: 7-9; April, 1972. (news)

Vander Ploeg, Jan. Copyright law and library photocopying: striking a balance between profit incentive and the free dissemination of research information. Indiana Law Journal, 48(3): 503-13; Spring, 1973.

An analysis of the increasingly divisive impact of reprography on publishing and libraries is presented. As the illsutrative case, Williams & Wilkins vs. the United States (NIH library and NLM) is discussed, especially the ramifications of the Commissioner's report vis-a-vis the legal arguments for both parties. Finally some of the proposed solutions to the photocopying problem are analyzed as to their acceptability to both the publishing industry and the libraries.

Wagner, Susan. Government loses photocopying suit. Publishers' Weekly, 201(9): 48-49; February 28, 1972. (news)

Wagner, Susan. Midnight oil burns after Williams & Wilkins rule. Publishers' Weekly, 202(7): 29-30; August 14, 1972. (news)

Wagner, Susan. Photocopying talks collapse; new solutions sought. Publishers' Weekly, 202(24): 20; December 11, 1972. (news)

Wagner, Susan. A visit with Williams & Wilkins. Publishers' Weekly, 203(14): 22-23; April 2, 1973. (news)

Wagner, Susan. "Fair use" carefully defined in copyright revision bill. Publishers' Weekly, 205(25): 31-32; June 24, 1974. (news)

Williams & Wilkins Company v. the United States: report of the Commissioner to the Court. Washington, D.C.: Court of Claims; February 16, 1972. 63p.

The Commissioner finds for the plaintiff, Williams & Wilkins, and presents his opinion in three parts: 1) a synopsis of pertinent facts (detailed facts are listed in Findings of Fact); 2) copyright law as it applies to the case in hand; and 3) other points raised during the case meriting comment. In part 2, the Commissioner discusses and attacks each of the points of the argument for the defendants, NIH and NLM, which he divides into five areas: 1) non-ownership of the copyright (defense claims author ownership); 2) real party in interest (two of the four journals brought in evidence for infringement by plaintiff are not owned by plaintiff but published under contract with two professional societies who do own the journals); 3) noninfringement (defense claims single copies not sufficient to incur liability and copying must include printing, reprinting, or publishing of multiple copies to be actionable); 4) fair use; and 5) licensing (work in four articles from the journals was supported by the Public Health Service therefore giving NIH and NLM license to copy).

<u>The Williams & Wilkins Company v. the United States</u> (decided November 27, 1973). Washington, D.C.: Court of Claims, no. 73-68; November, 1973. 104p.

The decision of the Court of Claims overturns the Commissioner's ruling in favor of the plaintiff, Williams & Wilkins, and finds for the defendants, NLM and NIH. The opinion is based on four major facets: 1) defendant's use of plaintiff's copyrighted material constitutes "fair" and not "unfair" use; 2) opinion is based on the "type and context of use by NIH and NLM"; 3) based on the records of this case, significant detriment to the publisher has not been found; 4) the present "judicial doctrine of 'fair use' is amorphous and open-ended" and courts should not take over the job of Congress in mediating compromises between conflicting interests. The decision emphasizes the uniqueness of the issue in court law and feels that to prohibit photocopying of the type on record would in the Court's opinion seriously harm medical research.

Williams & Wilkins: libraries win round one. <u>American Libraries</u>, 5(2): 60-61; February, 1974. (news)

Wilson, Samuel A., Jr. The United States Court of Claims on copyright: Williams & Wilkins Company v. the United States (487 F 2d 1345) in perspective. <u>Mercer Law Review</u>, 26(4): 1401-15; Summer, 1975.

The article discusses a history of the Court of Claims jurisdiction in regard to copyright, followed by an analysis of the Court of Claims decision in the Williams & Wilkins case, finding that defendants, NIH and NLM, were not guilty of infringement. He disputes the Court of Claims judgment ruling that the photocopying done in the respective institutions constituted fair use, arguing that it misconstrued and misapplied the intent of "fair use".

Yanow, Lilli A. Copying, the reconciliation of conflicting interests: an analysis of Williams & Wilkins Co. v. the United States (172 USPQ 670). Rutgers Journal of Computers and the Law, 3(2): 328-44; 1974.

This is an analysis of the expanded conflict between the needs of the public (including libraries, education and research institutions) for enlarging the dissemination of information and the interests of the author and the publisher, centered around reproduction of copyrighted works. A signal event in this conflict is the Williams & Wilkins case which is discussed with respect to the ramifications of this decision and its impact on the photocopying problem.